ALIEN ARTIFACTS

IS ANYONE ELSE ON THE MOON?

The Search for Extraterrestrial Evidence

VOLUME - 1

By
Ross Marshall

Is Anyone Else on the Moon?
"The Search for Alien Artifacts"
Volume – 1
By
R. S. Marshall

Moon Mars Monuments Madness

2ed Edition, c. 2015

Weirdvideos.com ISBN Numbers:
ISBN-13: 978-1515368076
ISBN-10: 1515368076

Cover Art credit by CreateSpace, Amazon Books
Interior book design by BEAUTeBOOK
Printed in Anacortes, WA USA

Weirdvideos.com c/o R. S. Marshall
P. O. Box 1191
Anacortes, WA 98221
www.weirdvideos.com

Ordering Information:
Quantity sales. Special discounts are available on quantity purchases by corporations, associations, and others. For details, contact the publisher at the addresses above.
Orders by U.S. trade bookstores and wholesalers. Please contact
Weirdvideos Distribution: Tel: (360) 421-7195;
or visit www.Weirdvideos.com

Printed in the United States

ALIEN ARTIFACTS

IS ANYONE ELSE ON THE MOON?

The Search for Extraterrestrial Evidence

VOLUME - 1

A Study in Selenography, whereby an attempt is made to determine whether there is evidence for extraterrestrial activity of space aliens. Illustrated with hundreds of NASA planetary pictures!

"When a person is honestly mistaken and hears the truth, they will either quit being mistaken, or they will cease to be honest."
-- Anon

Weirdvideos.com

Dedication

This book is dedicated to George Leonard, Fred Steckling, Richard Hoagland, William Saunders, Mark Carlotto, Mike Bara, Rob Shelsky, George Kemplani, Erich von Daniken, Zecharia Sitchin, David Childress, Josephus Goodavage, Jack Swaney, Felix Bach,

Don Ecker, Vito Saccheri, Rene Barnett, M. J. Craig, Victor Bertolaccini, Doug Turnbull, Tom Lehmann, Stephen Baxter, Ray Villard, Giorgio Tsoukalos, Stanley V. McDaniel and Monica Rix Paxson, Jason Martell, Philip Coppens, Louis Proud, Timothy Good, Alan McGregor, Stephen Webb, Maximillian de Lafayette, Whitley Strieber, Stanley McDaniel, Monica Paxson, Jim Marrs, Alex Milway, Adam Moon, Robin Moore, Xaviant Haze, Roc Hatfield, Mac Tonnies, George Haas, Brube Rux, Nick Redfern, Arthur C. Clarke, Gene Roddenberry and Lewis Carroll, with special thanks to Douglas Woodward.

CONTENTS

PREFACE

SECTION I
INTRODUCTION

SECTION 2
"IS ANYONE ELSE ON THE MOON?"
INTRODUCTION

PREFACE

"A truth's initial commotion is directly proportional to how deeply the lie was believed. When a well-packaged web of lies has been sold gradually to the masses over generations, the truth will seem utterly preposterous and its speaker, a lunatic." -- Dresden James

If you are reading this at the moment, then you must be one of those persons that is interested in space science. From the title of this book you must as well be intrigued to some extent with whether aliens exist. Furthermore, if you are considering that aliens do exist, then it is not that far removed from your mind to be considering that these aliens (if they exist) must have left evidence behind in their activities just as humans do.

The shape of one's eyes, whether they be human or olive shaped does not change the fact that all creatures leave effected trails. No matter where a living thing slithers, it changes the very cosmos around them. Even the simple observation of an object, in some way, changes the object being observed.

In quantum mechanics, which deals with very small objects, it is not possible to observe a system without changing the system, so the observer must be considered part of the system being observed. This is called the "uncertainty principle" and it has been frequently confused with the observer effect, evidently even by its originator, Werner Heisenberg.

The uncertainty principle in its standard form actually describes how precisely we may measure the position and momentum of a particle at the same time. If we increase the precision in measuring one quantity, we are forced to lose precision in measuring the other. An alternative version of the uncertainty principle, more in the spirit of an observer effect, fully accounts for the disturbance the observer has on a system and the error incurred, although this is not how the term "uncertainty principle" is most commonly used in practice (Wiktionary, under the term "observer effect").

Well, whether it be the "observer effect" or the "uncertainty principle", apparently, things get changed when living (observing) beings come in contact with them. Thus, it is without a doubt, that

.

if aliens have traversed the surfaces of our neighboring planetary systems and have done more than just "looked," we should find evidence. And, if they have dug
up the surface and built gigantic structures, colonial outposts, highways and tunnel systems, swimming pools, cafes and theaters, then we surly should find some evidence.

Needless to say, the "uncertainty principle" and especially the "observer effect" has been proved once and for all in the "observations" done by the following alien artifact hunters. The sterile lunar surface (as proved from lunar photographic sources and the regolith samples as obtained from our unmanned spacecraft and Apollo Missions) has been polluted with uncountable changes. The lunar surface is now riddled with (supposed) archaeological artifacts!

There is tons of lunar data to be studied and once we have collected the selected data we need, all we have to do is add it up to see whether the summation points to an alien presence or not. Furthermore, we must also ask just how much evidence is needed to prove one way or the other. If we find one conclusive piece of evidence, just one fragmentary artifact of non-human origin, then we need no further proof.

On the other hand, if we dig up both all of Mars and the whole of the Lunar surface and we find nothing, then even though it is still possible that aliens might exist elsewhere, it is most probable that they do not, at least on the Moon and Mars. In other words, one small "alien" step (footprint) would prove to us that man was not the first on the Moon. Alternately, the silence of one million billion tons of Martian and Lunar dirt sifted, observed, and classified as to bearing no evidence of alien activity, begs the possibility question and reduces the probability factor down to almost nothing.

Since, the most likely planetary systems that humans might be able to dwell upon are the Moon - being the closet, and Mars, the most potential planet to inhabit, unlike all the others, then we can expect to wonder if aliens might have done so long before us. After receiving pictures from the Moon and Mars missions, the question comes up as to what the many anomalies are that we find.

For example, the famous Face on Mars. Ever wonder what it is? Many people say this surface feature is evidence of alien activity! Others say it is a pile of rocks with a play of light and shadow giving an illusion of artificiality. Critics argued that light and shadow can play tricks on the mind, especially minds that are overly imaginative and educated. In other words, take for example, they say, where clouds often take the shapes of animals and human faces. The same is true of rock formations, such as the Great Stone Face supposedly on Mars.

Even before the latest Mars observation missions transmitted the Face's true constitution back to us, NASA scientists were telling us that it was just a blurry shadowy and highlighted mountain - a huge pile of rocks. This of course was based on the lack of evidence that it could be anything otherwise. The tendency of chaotic shapes to form patterns vaguely resembling familiar things is responsible for people making identifiable things out of anomalous things, and also the cause of many absurd books: rods, chemtrails, UFO's, ghosts, and other such nonsense.

Moreover, speaking of books and alien artifacts, the most absurd book ever written about Lunar and Martian alien artifacts is THE MONUMENTS OF MARS by Richard Hoagland, which we shall deal with later in this book. It may not be the funniest book ever written, but it is the most absurd one ever published. It is the most absurd only because it is the most well written and, unfortunately, the most dangerously tempting to believe.

More recently, UFO enthusiasts and mad Moon and Mars monument hunters have been playing 'find the hidden artifact' game with our planetary system. They pore over thousands of photographs of cratered surfaces until -- aha! -- they find something (questionable) suggesting the presence of outer space creatures. This is believable, if you have a vivid imagination, are a sucker for lies and can swallow the contorted, blurry, and twisted chunks of crater ejecta as unnatural artificial alien manufactures.

Notice, that before computers, it was all based on twisting and misinterpreting selenography; superimposing presuppositions over shady and overly highlighted lunar surface features and mistaking emulsions blobs and development splatters for alien artifacts. Since computers have come in, and since most of the aliens artifacts

3

based on previous research has been disproved as photographic processing and development mistakes, it is now jpeg compression creations from pixel perversions that are the alien artifacts.

As early as 1976, George Leonard, in *"Somebody Else Is on the Moon"* (Pub. by David McKay, 1976), carried this kind of speculation to such extremes that he managed to write one of the funniest books ever written on the subject. Leonard was an amateur astronomer and retired public-health official in Rockville, Maryland. In his book, George claimed that bridges on the moon were among the least controversial things about the moon. Really?

Well, needless to say, all bridges vanished when the Apollo photographs were obtained. The "bridges" were nothing more than illusions created by sun-light and shadows, yet this wonderful enlightenment did not stop the astro-myth spinners. The myth of alien artifacts and structures still persists in alien presence based literatures to this day.

The same thing happened to the mysterious "spires" on the moon. Photos in 1966 of the Lunar surface showed objects casting such long shadows that alien artifact hunters decided they had to be rocket ships or radio beacons -- at least "something" built by aliens. A Russian periodical called *Technology and Youth* featured a wild article about these spires in its May 1968 issue. The spires turned out to be ordinary boulders, because it was found that their long shadows were caused by the sunlight hitting them at extremely low angles. If it was not Clementine's new pictures exposing the alien artifact hoax, it is now the Lunar Reconnaissance Orbiter. No more can we say "Somebody Else is on the Moon" as we shall see.

Leonard quotes an unnamed NASA scientist in his interview in the tabloid *Midnight* (February 8, 1977): "*A lot of people at the top are scared.*" He thinks the aliens live underground and that seismic quakes on the moon are caused by their undersurface activities. He said, "*NASA is simply lying to the American people about UFOs,*". He suspects the aliens are waiting patiently to take over the earth after we blow ourselves up. Photos of the Moon's surface, he insisted, show rims of craters sliced away by giant machines, jets of soil spraying out (caused by mining operations), and tracks of huge vehicles. "*No, I do not know who they are,*" Leonard said, or "*where they come from or precisely what their purpose is. But I do know the government is suppressing the discovery from the American people."*

Really? Have we ever seen a detailed close-up of these areas? The Lunar Reconnaissance Orbiter is now showing us "in detail" in each magazine of pictures that there is nothing more than lunar dirt on the Moon with no hint of alien activity.

In 1981, Fred Steckling published what is to be known as the second best funny-book ever written on Moon artifacts – "We Discovered Alien Bases On The Moon." Fred pawns off the same nonsense as George Leonard: Alien ponds, lakes, digging machines, pipes, tubes, screws, and oil tank fields, crater rim space craft landing pads, space craft hangers, and on and on. It is still published by the

Adamski Foundation.

Seeing familiar anomalies on Mars has been common ever since the invention of the telescope. Percival Lowell found the red planet's surface so honeycombed with canals that he wrote three books about how the Martians, desperately in need of water, built the canals to bring water from polar regions. Now, of course, we know the canals were only a figment in Lowell's mind, distinguished astronomer though he was.

Unfortunately, this has not deterred seemingly intelligent people from similar self-deception. Even though such canals were later proved to be illusions derived from the play of shadows and the sun light reflecting off the dust storms on the Martian surface, Martian Oopart hunters continue to make alien mountain motels out of crater ejecta hills.

Here and there on Mars are formations with grid-like structures. "Did NASA Photograph Ruins of an Ancient City on Mars?" is the headline of a *National Enquirer* article (October 25, 1977). A photo of a region near the Martian south pole shows a series of square-like formations called "Inca City" because they somewhat resemble a decayed Indian village.

In 1977, electrical engineer Vincent DiPietro came across a 1976 photograph taken by the Viking spacecraft that orbited Mars. At first he thought it was a hoax. The photograph showed a remarkably human-looking stone face about a mile wide. NASA had released the photo shortly after it was taken in 1976, and

5

planetary scientists emphasized that it was a natural formation. DiPietro thinks it isn't. Computer scientist Gregory Molenaar used image-enhancement to explore details of the face, and in 1982 DiPietro and Molenaar published a 77-page funny book, *"Unusual Martian Surface Features,"* about their results

The authors concede that the face may have been produced by erosion but they suspect otherwise. They claim that computer enhancement shows an eyeball in the Face's right eye cavity, with a pupil near the center, and what looks like a teardrop below the eye. *"If this object was a natural formation,"* they write, *"the amount of detail makes Nature herself a very intelligent being."* The more contemporary Richard Hoagland, if I remember correctly, believes that there are teeth to be found in the Face's mouth peeking through the shadows.

West of the big stone face, in the shadow of a pyramid-like formation, is a grid-like pattern suggesting a lost city with an avenue leading toward the face. (See "Metropolis on Mars," an unsigned article in *Omni*, March 1985). Skeptics have pointed out that the so-called pyramid is much cruder than scores of pyramids found as natural rock formations in Arizona. But no, the alien propagandists are so compelling in their arguments that even Hollywood was convinced enough to make a movie in support of this fiction - "Mission To Mars." In this fictional account trip to Mars, the Face turns out to be an ancient alien UFO hanger, built eons before mankind ever existed. The fiction of course supports and propagates evolution by making the aliens the forefathers of mankind and the creators of the apes we came from.

Top drum-beater for the view that the stone face proves that an alien race once flourished on Mars is writer Richard Hoagland. He has published a book called The MONUMENTS OF MARS which tries to prove this. There are mathematical calculations all through, with descriptions, myths, inter-dimensional geometry and plenty of Martian topographical language to authenticate it.

The Mars Face, Pyramid and City of Cydonia Complex are especially important in his book. Even though the latest Martian pictures reveal this Face to be just a piles of rocks - and Mr. Hoagland should be happy that NASA fulfilled his dream of obtaining these better images - Richard continues to insist that it is an alien sculpture and that NASA has tampered with the new images!

Fred Golden, writing the "Skeptical Eye" page in *Discover* (April 1985), ridiculed Hoagland's claims and ran a photo of another spot on Mars, where the topography resembles Kermit the Frog. Many other Martianologists have found, and are still finding, what they believe are anomalous objects of alien origin.

The Internet now is teaming with wild and crazy web artifact web sites. It has apparently come to the point of pure competition as to who can dream up the best nonsense and with no care at all as to validity or factuality. Many of the blogs and web sites set rules to disallow negative and critical postings – all posts must support the competition.

If you search any kind of chaotic data, it is easy to find combinations that seem remarkable. Every page of a book of random numbers contains patterns with enormous odds against them if you were to specify the pattern before generating the random numbers. Every bridge hand you are dealt would be a stupendous miracle if you had written down its exact pattern before the deck was shuffled. But no one seems to have done either.

Furthermore, about every 20 years all previous pro-alien speculation upon the data collected prove to be just that - "speculation." Nevertheless, with new technology, new and more exciting "facts" are manufactured and pumped out of the spin doctor factories in support of the alien presence. Thank (the computer) god we now have pixel distortion!

Let someone close his eyes and talk for 15 minutes about a scene he imagines. You'll have no trouble finding amazing correlations between his description and any randomly selected scenic spot. Let a psychic crime-solver rattle on for an hour about clues to a missing corpse. It's inevitable he/she will have made some lucky hits if and when the body is found. Now, give a group of artifact hunters a planet with disturbed surface features and inevitably they too will shovel-up "evidence" of the existence of aliens. This is unbelievable but it is very explainable why they do what they do.

Close your eyes and thumb-press your retinas, and you will see all sorts of things, and every other imaginable thing, other than real tangible things. For, the creative human imagination is a wonderful device, but when used to create illusions to pawn off as reality, and make a buck ta' boot, that imagination is a con-artist.

This way of seeing "alien" things and selling the stooge a pro-alien artifact book for a fast buck is not too far removed from the ancient (extortion) arts of Tesseography, haruspicy, extispicy, hepatoscopy or hepatomancy and kypomancy (divination by tea leaves, animal guts and coffee grounds) and especially, when dealing with the geologic features and materials of mountains, sand, rock and dust - oromancy, abacomancy, aichmomancy, amathomancy. No matter what the lunamantic diviner sees in moon rocks, cracks, shadows and highlights, the poor under-developed phren believes it, pays up and toddles off with his newly purchased enchantment.

The great and acclaimed raving chresmomancist Richard Hoagland has mastered the shadowy arts of the ancients, taking the royal position as Vicar of alien artifact morpho-divination – the new selenomantic of lunar research. With exoenthusiasm and the calligraphic whisk of chartographomancy, the new Hoaglandomantic gut-busting geloscopeable topomological science was born, published and dumped on the market, and the green-backs roll in as the new pseudomology spreads like wildfire within society.

The list of alien artifact con artists, entertainers, pabulum producers, paradigm programmers, pseudoscientists and pseudomantics is practically endless. The alien artifact pulp rags continue to pile up even all the way to the moon! If the cow never made it, Hoagland's book "Dark Mission" surly did.

The publishers range from blind pseudo-scientific writers such with Leonard, Steckling, Hoagland, David Childress, Erick von Daniken, Sitchin all the way to pure science fiction writers and game designers like Doug Turnbull, Tom Lehmann and Stephen Baxter. The first believe their own fictions, while the latter wish their fictions were believable. Nevertheless, in both cases, the idea of space intruders is advanced at the expense of truth.

Some last words on these people must be noted. In almost every case, it will be noticed that these rags start with the presupposition (false premise, idea) that aliens exist. From this porcus-swill all data is bent to support such nonsense. This is the easiest way to pawn off a fast buck-maker. To approach it any other way is time consuming, tedious, technical and requires much source tracing and testing. This is the preferred way used in this book – start with the data and work towards a conclusion.

There are many reasons why pseudo-science books are written the way they are. First is to sell the book on a large scale quickly. Secondly, to reinforce the idea that aliens exist and lastly, to replace good-ole' scientific and religious truth. There are two witnesses against alien presence: Sacred Scriptures and government space agencies. Science has come to the conclusion that the probability of finding evidence for the existence of aliens is zilch!

They have not ruled out the possibility, but the probability has decreased tremendously. Religions have, alternately, pointed this out by defining the probability factor as being non-physical, spiritual beings called good and bad or unfallen and fallen angels. The religious view more closely supports the hyper-dimensional interpretation.

In his book we will not debate whether aliens really exist. This writer actually believes that "something" alien to us exists, but proof is lacking as to their physicality. What will be analyzed is whether what evidence we do have supports the theory that aliens have visited our solar system. It is most probable that my readers have waffled through other alien artifact books supporting the alien presence. It is now time for alienomantic enthusiasts to hear the other side of the argument.

This writer once believed "We Never Went to the Moon" (Ref: Bill Kaysing) only to find out later that we probably never left. Recently the facts have dictated that we did go and that there is no real need to go back but for a few more samplings or until technology advances further to support human occupation.

The next deception came with "The Face on Mars." Mr. Hoagland and Carlotto did a wonderful job in convincing the public of an alien presence on Mars. But no one was happy with the blurred pictures. It was not until later, after further Martian spacecraft photography that the Hoaglandite belief came tumbling down into a pile of rubble. The face was only a misinterpreted heap

9

of rocks. But, this did not stop the con-game of pawning off the theory of E. T. intruders.

As soon as the more recent Martian pictures came in, and quickly after the Mars face was debunked, thousands of new artifacts popped up. With the new LROC and the exposure of the moon to higher resolution photography, the public eye quickly shifted from lunar to Martian observations. But not with this writer. I soon realized that the redirecting of the public from lunar searches – since there was nothing to be found there, one could continue selling the gullible the same belief system. As moon photography was to Leonard, Steckling, so the Martian photography is to us today.

Now, it is believed by this writer, as the lunar pictures are to us today – sterile of any evidence of alien presence - so one day will be the Martian photography. Mars will be another study as the photography gets better. For now, let us see what the higher resolution LROC photography has revealed about aliens on the moon and what science has determined about interstellar space travel. The likelihood that aliens have visited us from another star system appears to be highly improbable. In the process of collecting data keep track of all the "zilches!"

SECTION-I

INTRODUCTION
PLURALITY OF INHABITED WORLDS DECEPTION

"Because false doctrine is so abundant: "The Truth will be the Strangest Thing you will ever Hear" - **Paul Smith**

The belief in the plurality of inhabited worlds, which for centuries has had its advocates among famous authors and philosophers, took deep root in many cultured minds several generations ago. During the last 50 years, however, progress in astrophysics, biology, planetary studies, and [more recently], satellite planetary exploration eliminated, first of all, the stars as a whole, and then nearly all the planets and their satellites as possible abodes of life, as we know it on earth.

Nevertheless, in recent times, a rage of E.T. artifact madness has taken a strong grip upon a large segment of man. Hence, opinion has split into two opposing camps, both arguing from a more or less well founded metaphysical basis, and using the same accumulated space science data for supporting their positions. Evidence seems to be plentiful, but irrefutable conclusive proofs are hard to find.

The first points out the exaggerations to which belief in life on other planetary bodies has given rise. The Second, on the contrary, stresses the amazing variety of life forms on Earth, and the powers of adaptation, and is ready to extend this existence as a kind of 'generalized' life, in which life on Earth would only be a 'special' case.

As to which one is correct, the first makes the mistake of assuming ANY and all life, [if it exists at all in space and on OTHER planets,] must fit within the narrow limits of life giving properties, as Earth defines them for humanity. This view derives its conclusions primarily from physics, astrophysical researches, and planetary probes. The second commits an equal mistake of presupposing the "universality of life", assuming no minimal conditions for its manifestation.

This view derives its conclusions mostly from visual observations. The first denies all but carbon based creatures, while the latter extends the basis of life to silicon and other types, even to extra-dimensional life forms.

11

Obviously, each are completely objectionable to the other, yet both share some commonalities. They both are mutually antagonistic, and both fall victim to an over emphasized non-revisable presupposition. In any case, the only field where these ideas could possibly be confronted with observational and experimental data, is that of Planetary Physical Astronomy.

In recent years, the only planetary bodies to be pin-pointed as probable locations for ET artifacts are the Moon and Mars. And this brings us to this research presentation. To discuss the evidences as to whether there is any ALIEN ACTIVITY on other planetary bodies, other than Earth.

It is NOT the intent of this study to re-open the discussion in the above extreme, but, on the contrary to try to present materials to clarify the matter. This book is not meant to confuse it, as many fellows have done, but rather to simplify, and bring the reader back to a fresh and unbiased mind to the FACTS, as to whether the evidences found point to E.T. activity on our planetary neighbors.

Our approach is simple and scientific and is as follows: 1.) We must return to the original sources, where possible. 2.) We must use what is authentic and discard what is erroneous. 3.) We must properly interpret the area of study and gain the TRUTH as to its nature, and discard what is obviously erroneous; or to conclude a neutrality, until further research can be obtained to decide, [without a SHADOW of doubt], whether it is NATURAL or ARTIFICIALLY created. 4.)   We must present the FACTS and evidences to the viewer with what we have found to be true [or, at least possible], and then allow the viewer to decide for himself. 5.) We must supply the viewer with resource connections and outlets for the viewer's personal pursuit of the matter. Only this way can we come to terms with who is blowing Lunar dust and who is not.

The above numbers 2 and 3 are of extra importance, when dealing with data. The question arises as to what is authentic and conclusive and what is not and is anomalous. And how do we go about deciding the difference to come to a conclusive "conclusion," that there are truly alien artifacts on the Moon?

I would like to present a mathematical formula. Let us take a set of data, say 100 pieces of material (in this case, an analogy for

NASA data, photographic materials, eye witness testimonies, etc.). Let us "tag" each item with a positive or a negative symbol. The *positive* symbol is to refer to those pieces of data as conclusively "of non-human origin" or "artificially created by non-humans."

The negative symbol is to refer to all those pieces of data that are non-conclusive of the above. Now, let's take that set of 100 pieces of data (try maybe 1 million pieces?), for example, 100 photographs of the Lunar surface . If each piece is decided to be non-conclusive as to supporting alien presence, we have 100 negatives. Let us ADD up all 100 negatives. Do we get a positive answer? Does -1, plus -1, plus -1 = equal +3? The answer is no.

Now, let us do this with all 100 negative pieces. You will get -100. Adding up 100 inconclusive pieces of data gets us a big fat negative pile of inconclusiveness. How then can people say that all these negatives add up to proof that aliens are on the Moon?

The opposite result would get us 100 positives, meaning we have 100 totally conclusive pieces of data telling us that there are aliens, and that they have been, and maybe still are on the Moon. Hell,... even ONE positive piece of data would be enough! Should we say any more? Please, can someone bring forth just ONE positive piece of evidence?

For centuries, the planet Mars and the Moon, have been the "FIRE" in many planetary observers. Many have devoted their whole lives to such studies and have even built observatories to delve deeper into the many unanswered questions. Along this path of planet watchers we will find many wild theories and tales to read.

If we apply the above rules, we will determine quickly the difference between falsity and truth, and also whether there is any alien presence or "life" extant in our planetary system other than on planet earth.

When it comes to data, resources, artifacts and testimonies, one will find it is no easy task. There is so much emotional and bias thinking, as well as outright deception, that even the best scientist can be fooled. Astro-archaeology is as speculative as literary archaeology, and can sometimes be a quandary as to deciding what is real and what is poetic and/or fictional.

A good example of confusion in decision making over, for instance, ancient claims and testimonies of extraterrestrial activity, or at least ancient advanced human technological achievements can be sampled in the following.

The ancients themselves "claimed" many, surprisingly modern sounding stories of space travel and heavenly contacts. The ancient Hindu's, in their astronomy book the Surya Siddhanta ["The God Surya's Astronomy," 500-2000 BC] claim that their older ancestors went to "MAYA" the Moon and that the "Siddhas" and "Vidvaharas," or philosophers and scientists, had the ability to "examine the regions below the moon but that are above the clouds" They found [2000-4000. BC?] that the Moon had NO AIR, was DRY as powder, and looked as if it was covered with *"Burnt Ash and Dust"* (The Surya Siddhanta. See Andrew Tomas, "We Are Not the First", London: Sphere, 1971, p. 149).

In the 3rd century B.C., Chuang Tzu, in a work entitled "Travel to the Infinite," relates a trip the Chinese made into space some 32,500 miles from the earth. The following is a synopsis: In the year 2309 B.C. the engineer of Emperor Yao "Hou Yih" decided to go to the moon. The *"celestial bird"* provided Hou Yih with information on his trip. He explored space by *"mounting the current of luminous air"* (i.e. the exhaust of a fiery rocket). Hou Yih flew into space where *"he did not perceive the rotary movement of the sun."*

On the moon he saw the *'frozen-looking horizon"* and erected a 7 STORY building, which they called *"the Palace of Cold"*. (This statement is of some importance in corroborating the story because it is only in space that man cannot see the sun rise or set.) His wife Chang Ngo likewise made a trip to the moon, which she ACCURATELY DESCRIBES as a *"luminous sphere, shining like glass, of enormous size and very cold; the light of the moon has its birth in the sun,"* she declared.

Chang Ngo's moon exploration report appears to be surprisingly correct.

The Apollo Mission astronauts found the moon desolate with a glasslike soil with parts of it even paved with pieces of glass - probably melted from the intense heat of the Sun. Most of the Moon, at any given time, is in the extreme cold. It plunges to minus

14

250 degrees Fahrenheit at midnight. But in the day time, it swings to just the opposite, 250 degrees above. That's hot! No need to argue whether the Moon can be hot and dry as powder. Not to digress to far here, but notice that none of these accounts mention any alien presence or extraterrestrial threats.

In Tibet and Mongolia, ancient Buddhist books speak of *"iron serpents which devour space with fire and smoke, reaching as far as the distant stars."* The Epic of Etana (4,700 years old) supplies us with very accurate descriptions of the earth's surface from progressive altitudes – descriptions which were not verified in our own era until the high-altitude aerial flights of the 1950s and the first space shots of the 1960s.

The description of this ancient space flight depicts exactly what happens when man leaves the earth (the concept of the round earth which becomes small, due to perspective as distance increases, and changes into particular colors). The ancient Book of Enoch said that in space *"it was hot as fire and cold as ice"* (where objects get hot on the side illuminated by the sun and icy cold on the shaded side) in the "dark abyss."

Yes, these claims are very suspicious of a highly advanced technology lost in past times. Mythology claims even that the ancients built Moon bases and cities, then blew them up. Hence the lunar glass dome theory and the atomic bomb theory for lunar crater formation! For, after all - *"The Gods and demons behold the Sun, after it is once risen, for half a year; the Fathers (pitaras), who have their station in the Moon, for half a month;"* (Text Book of Hindu Astronomy. xii-74. F. Burgess & W. D. Whitney).

On the other hand, these stories could be no more accurately depicting the historicity of ancient space travel or alien activities than the works of: Cyrano de Bergerac, in his "L'autre Monde Ou Les Etats et Empires De La Lune" 1657. - *"Here is how I betook myself to heaven,"* Bergerac writes. *"I attached to myself a number of bottles of dew, and the heat of the sun, which attracted it, drew me so high that I finally emerged above the highest clouds. But the sun's attraction of the dew drew me upwards so rapidly that instead of approaching the Moon, as I intended, I seemed to be farther from it than when I started. I broke open some of the bottles*

15

and felt my weight overcome the attraction and bring me back towards the earth."

Or, Baron Karl Friedrich Hieronymus Münchausen, in his "The Surprising Adventures of Baron Münchausen" 1785. - Like de Bergerac, the Baron was both an historic and literary figure. In life, the baron was a German nobleman. Yet, in the pubs, the notable Baron was a story teller of tall tales: For, the Baron *"rode a cannonball," "danced in the belly of a whale,"* and *"traveled to the moon"* – and he did it twice! The famous "story" tells us he did it with multiple strips of bacon tied to a long rope, swallowed by (so many) geese, and with the blast of a pistol, off he went.

Or, Jules Verne, in his famous "From the Earth to the Moon" 1865. - Jules Verne's humorous science fantasy begins when members of the Baltimore Gun Club devise a plan to manufacture a giant cannon that will shoot a "space-bullet" from Florida to the Moon. Verne deserves applause for his visionary calculations.

When the Moon landings were finally achieved nearly 100 years later, the calculations from his story were startlingly close to modern scientific findings and, last but not the least; H.G. Wells, in his fabulous "The First Men in the Moon" 1901. In this tall tale of scientific sounding (at that time) fiction, we see impoverished Englishman Mr. Bedford pair up with Dr. Cavor, a scientist working on a gravity-shielding material he calls 'cavorite'. After discovering a peculiarity in the material's behavior that makes the air above it weightless, the men build a spherical spaceship and travel to the moon where they are captured by insectoid moon-creatures. Well, as we reach the 1930's and 40's, this tradition of fictional story-telling increases until every science fiction book, comic strip, and (especially) the movie industry contained many parallels to modern space travel. Apparently, it is not all that impossible to foretell technological advances ahead of time as the imagination is a wonderfully creative machine.

Because of the close proximity of the Moon to the Earth, Governments today - just like the ancients of old - are infatuated and addicted to the mysterious world between the stars. Look at the world wide efforts in building scientific space craft.

Attempts to reach the stars dates back centuries, with one hilarious example being a Chinese Emperor who blew himself to smithereens from attaching solid fuel rickets to his royal bed figuring the engines would carry him to the stars. He did make it somewhere and we hope he made it into heaven but there is no proof he ever made it to the stars.

It is no wonder that many persons and Governments have a permanently fixed vested interest in space travel, colonization and now, radical support for ET occupation of the solar system. Without the "mythos" as a fire, there is no strong reason for the masses of humanity to pursue science, other than to escape the planet earth. Raw science is somewhat dead to the common man. Only a true scientist can extract for other scientists, sweetness from the sour grapes of EMPTY LIFELESS SPACE - and that is exactly what EMPTY LIFELESS space is it all about for billions and billions of light years distance: Nothing, just lifeless space!

Nevertheless, these are no sweet grapes to the mythologists: They reason "*There MUST be ALIENS out there. Or, we don't care about Space Science! What's it to us?.. We'll never go there! So, why should we waste our precious time (and money) on unimportant (unprofitable) things?*" So, if we are to invest into it, we will have to stick a few aliens in space, plant a few monuments on the Moon, and people will get interested. Why? Because NOW the masses have HOPE, whether they ever go for a space ride or not.

Could the ALIEN FACTOR be just an alternative HOPE mechanism to traditional religion? Could the argument for ALIEN technology be just a convenient alternative to Human technology, which seems to be 200 years behind getting us all off this DYING PLANET? Could the subconscious desire be that, since humanity cannot get off the planet, then ALIENS can do if for us? "*Let's make a deal with'em!*" Maybe this is another topic for another book investigation? Or, maybe, because without some E.T. mythos, the Government (that coincidentally, neither cares to prove or debunk the urban legend) would not have near the publicity and funding it does.

Before the advent of artificial probes to the moons and planets of our Solar System, in the 1960's, people had only the benefit of Earth bound observations. The opinions were the same, though. Some were adamant about E.T existence on other planetary worlds, while the others were of the conservative belief that, if there is, it is of a more primitive type, like say, ORGANISIMS, or LEACHENS. Some went so far as to allow, mosses and plants and grasses to exist on certain locations of the Moon. Even before this time, others were digging canals on Mars, until they were proved to be changes in atmosphere caused by huge dust storms. (We may forgive these people for the lack of technology to prove otherwise).

For example, in the 1700's, Emanuel Swedenborg claimed to have ascended to the planets and found that ALL the planets had life on them. The Moon even had huge TALL Quaker-Like people on it. Now, this guy was the first to theorize the nebula theory, was an inventor, and a very smart man in many fields. Yet, I guess for pleasure (and profit?), he pawned off the alien theory to promote the sale of his products as well as his religion.

In this century, supporters have varied from the wild claims of ET fanatics, to the scientific Government opposition of the E.T. presence. In the last century and in the early part of this century, many opposing debates sounded throughout the scientific world. Astronomers such as Dr. Percival Lowell propounded that life does exist on other planets, like MARS. He *believed* and tried to prove that Mars had irrigation canals, green "seasonal" plant life growths, canal "pumping stations" and finally, of course, a highly advanced race of intelligent Martians living on the planet. Naturally, this led to such Sci-Fi entertainment's as "The War Of The Worlds." The brilliant Orsen Wells had a blast deceiving the public - all the while he stuffed his gut full with the proceeds! As the decade went along, Hollywood even went so far as to put a Robinson Family into Space.

Nevertheless, Lowell had his opponents, such as Alfred Russell Wallace in, "Is MARS Inhabitable?" (Pub., 1907), who proves him to be more imaginative than scientific. Referring to Lowell's fanatic Martian myth making, he said "...*that animal life, especially in its higher forms, cannot exist on the planet. Mars, therefore,* [After 110 pages of arguments] *is not only uninhabited by intelligent beings such as Mr. Lowell postulates, but is absolutely UNINHABITABLE.*"

Now, whether it is uninhabitable or not, we'll leave it up to future colonists to decide. As to the evidences of Alien occupation, well, this is what we are centering upon. CAN WE FIND ANY evidence of Alien Activity? Patrick More in his, "The Planet Mars" (1950), rather suggests, after 90 pages of arguments, that *"Without speaking of the 'Martians', we cannot therefore say that there is or is not life on Mars; but we can say,...that the immense majority,... would find it impossible to accommodate themselves to the conditions, which prevail on Mars. In particular, we ourselves could not live there in the open air."*

Mr. More obviously means, Mars is INHABITABLE, if we artificially accommodate ourselves to the planet's environment. A hidden desire to believe, that if there are ET's, that they might inhabit Mars under artificial conditions? Mr. More hasn't said this, but the door is left open. And wide open it is, for in the classic 1955 book, "There IS Life On Mars", Mr. Nelson begins, what has turned into a full-fledged alien infestation of our Solar System. He wedges the door open by saying, *"It is most unlikely that intelligent life exists there [on Mars], although even this is not an impossibility...Even IF intelligent creatures did exist,...they would probably bear no remote resemblance to us..."*

As a conservative scientific alternative to this open door, NASA, in their publication "The Book of Mars" (NASA SP-179. 1968 Edition, Ref.1), had concluded from the Mariner probes, that, *"In view of the almost complete lack of oxygen and the absence of any significant quantity of liquid water, it is highly improbable that there are now any advanced life forms on Mars."* Such statements about Mars coincided with what we were learning about the Moon too. Up to 1969, nothing had landed on the Moon but artificial satellites. The Government had only landed probes and orbited the Moon taking photographs of the surface. Just as Mars was to be photographed later, the Moon had been under the photographic eyes of the Ranger, Surveyor and Lunar Orbiter Missions, and the preliminary Lunar orbits of the Apollo Mission. Many photographs were taken. The numbers range into the 10's of 1000's of frames. Unfortunately for alien artifact hunters, there is not one shred of conclusive evidence to alien presence to be found.

As to the silence of Government on the issue, the myth spinners say, the Government is erasing all evidence, only because proof of alien presence would destroy human civilization. This is bunk! Over half of humanity "believes" in aliens! The truth spinners say, actually, the reason why Government does not come right out and fully debunk alien presence, but let the Hoaglandites rattle off the *bull*, is because the circus barkers bring them lots of votes and money to boot! Thanks to the World Wide Web, NASA has grown into a vast scientific social network unlike the dry laboratories of the 60's and 70's. It was during THIS period, from the 1960's up to the present, that the specialized field of ALIEN presence on planetary bodies really exploded into what it is. Notice also the financial support for NASA and other government space programs increased too. A few writers grabbed the NASA materials and began their studies to prove ALIENS were occupying the Moon. A scientific follow-up to the Hollywood manufacturing going on with Lost In Space, Star Trek, and the likes. ALL Fabian Socialist; all New World Order converts, and all Alien Agenda propagandists!

It was only a few years later that Mars would be dragged back into the Lowellian sphere of the "occupied Mars" syndrome. All the conservative kicking and screaming in the world, would do no good. Neither would the conspiratorial arguments, such as that of Bill Kaysing, who claims "We Never Went To The Moon," do one thing to stem the tide of the Moon Mars monuments madness. When ignorance rules the minds of audiences and laziness rules their ability to reach out and open a NASA book, people like Leonard, Steckling and Hoagland will always get away with pawning off mythos as science - there is a grand pay check involved here!

The Moon had already been established as a base for aliens, according to George Leonard and Fred Steckling. Now Mars was to undergo another round of treatment. If it wasn't Fred Steckling "sticking" us with George Leonard's older nonsense of "Someone Else Is On The Moon," it is now Richard Hoagland hosing off the old Lowellian dream of Martian cities and canals, and with Egyptoid FACES and Sphinx to boot! And, no doubt eventually, some hieroglyphics popped up as proof, as were supposedly found on Roswell UFO Crash metals.

Roswell Crash Metals? Now, is there alien evidences on our planet and the Moon? Has the Government sinisterly covered up all

20

the "facts" that would prove such existence? With all the studies and claims floating around, WHO IS HOAXING WHO? Who is denying the other the truth? Who is falsifying and who is really doing the fact-finding?

Yes, we should all agree that a man has to eat and make a living, and if he writes a book to do it, that's O.K. If he produces a video, and tells a story, that's O.K...and, if he cooks for a living, that's O.K. too. But, ...IF I go and order a "real" hamburger from a vendor, I expect a real one out of the oven, not the plastic one in the glass show case. There are real hamburgers and there are rubber "sodium-nitrate" ones.

However, science and technology has increased so far, that sometimes the fake is confused for the real, so much so, that sometimes even the cook does not know the difference. Scholastic bowel troubles and stomach problems? Yes. It has gotten so bad that it hurts to dig into the situation and really find out. Nevertheless, if you are going to really get your stomach full, you have to bite into it and eat it to find out. So, let us MUNCH OUT on MOON MARS MONUMENT MADDNESS and see how eatable it is, or rather how reliable the information is.

Let us find out whether the supporting evidence actually supports the theory of alien occupation and habitation of our neighboring planetary orbs.

There are a number of resources to consider when we pursue this subject, and NASA has tons of it to buy at a cheap price. There are articles on crater development, lunar morphology, selenology, volcanism, dirt, rock, soil and meteorite thin slice studies, and Lunar Transit Phenomenon, [LTP's: Ref.3], which are plentiful. This last source of anomalies dates as far back in historical times [not to exceed beyond the 1700's] as the late 1700's; anomalies that resemble very new Lunar Transient evens, such as an object passing before the Moon.

This object, by the way, was recorded by the Apollo-15 Mission. [1] What the Hell is it? And, who does it belong to? No earth government has claimed it yet. this makes you wonder if it might be alien in origin. This is a real fact to interpret and play around with. Could it be an orbiting alien spaceship? What about it being an orbiting fragment of rock? Look it up on "YouTube" and decide for yourself. [1a]

The Moon has been, for a long time, an object of Mystery. Early Moon watchers claim to have seen many wondrous things,... things that only today seem to easily fit into the ALIEN PRESENCE THEME. To them it was a little different. They had no answers, even though Swedenborg would have, and probably did, tell them that, what they were seeing were the lunatic religious Quakers. One early eye witness in 1790, a Mr. Frederick Herschel, was looking through his telescope and saw, in time of a total eclipse, "*many bright and luminous points, small and round.*"

In 1794, Dr. William Wilkins of Norwich, was amazed to see a light like a star appear on the dark disk of the Moon. He said, "*This light spot was far distant from the lighted part of the Moon...it lasted for 15 minutes...was fixed...(and) brightened. It was brighter than any light part of the Moon, and the moment before it disappeared, the brightness increased.*"[Ref.4]

Others from different succeeding periods saw many other things to add to the list of Lunar Transient Phenomenon. In 1824, Oct. Mr. Gruithuisen of Holland, a Selenographer, saw a light on the dark part of the Moon. In 1825, on Jan.22, two British Officers of the H.M.S. Coronation saw in the Crater Aristarchus, a light project from the Moon's upper limb and vanish.

In 1867, Thomas Elger claims he saw a flaming star-like light flash out from the dark part of the Moon. He said, he had seen lights on the Moon before, claiming this one was the brightest he had ever seen. He further stated, in his report to the Astronomical Register, that there were funny objects near Craters Birt, Sword shapes, crosses, and geometric shapes around other craters. In 1870 Lights were seen in the Crater Plato by English Observers.

In 1877, Monsieur Trouvelot, of the Observatory, near Paris, saw in Crater Eudoxus, a fine luminous LINE or cable drawn across the crater. Others in 1877 saw other strange things: Mr. Barrett saw a bright light inside Crater Proclus; Mr. Harrison, in New York, saw a light on the dark part of the Moon. A contemporary of his, a Mr. Dennett, affirmed this too.; Mr. Klein reported to his French government, that he saw a luminous triangle of lights appear on the floor of Crater Plato, they were followed by others which flowed across to the Crater Plato.

Between 1885 and 1919, many other sites were seen. LTP's such as lights, reddish smoke, curved objects, illuminated orbs, black areas whitened, luminous cables, black spots with white borders, floating spots, red shadows, dark round objects floating around the Moon's Face, dazzling white flares and light explosions. The reports are recorded into the 100's by prominent people, not to mention all the commoners who were never recorded. IS this evidence for ALIEN activity of the Moon?

Some possible explanations for LTP's can be found in a little publication called "LUNAR TRANSIENT PHENOMENA, by Sky & Telescope, March 1991. They list the following as possible causes for such LTP's. But, this list does not answer all the LTP's observed. LTP's can be caused by one or a combination of: Tidal Forces, Albedo Changes, Thermal Shock, Magnetic Solar Plasma, Ultraviolet Radiation, Solar Winds, Spectral Diffraction's, Meteoric Strikes, Moon Quakes, False Color and Piezoelectric Effects.

In more recent times, writers and researchers have advanced this conjecture even to the point of providing photographic proof for the existence of ALIEN ACTIVITIES on the Moon and Mars. Josephus Goodavage, George Leonard, and Fred Steckling are some. Others who have followed are Mr. Swanev, Felix Bach, Don Ecker, Vito Saccheri, Rene Barnett, David Childress, Mike Bara and last but not least, Mr. Richard Hoagland.

All these and many more upcoming Moon Monument theorists propound the ALIEN OCCUPATION AND INHABITANCE theory of our Moon and Mars. We must wait still, before we see Venus and the other planets become added to the list. Emanuel Swedenborg would be proud to see this day as he propounded (ignorantly?) that all the planets housed intelligent life.

To get back to the 1000's of NASA photos that supposedly show alien occupation of the Moon and Mars and the tons of comic books written on the subject, we come to the first excellently done comical study on MOON MONUMENTS. Mr. George Leonard's work "SOMEBODY ELSE IS ON THE MOON."

23

Taking Mr. Leonard's and Mr. T. H. Huxley's advice, which is quoted by Leonard, in the front of his book and with added interpretations, let us, *"Sit down before fact as a little child, be prepared to <u>give up every preconceived notion</u>, follow humbly wherever and to whatever abysses nature leads, or you shall learn nothing."*

Let us also take this advice and apply ourselves to the re-evaluation of Mr. Leonard's lunar lunacy; to Mr. Fred Steckling's photographic phantasms, and move on to a few more - (chuckle) - "lunatics", such as the monumental revelator, Richard Hoagland and his mountain of Moon Mars misadventures.

Our search to find whether there are aliens in our solar system and to verify if there are E.T. artifacts on the Moon as well as Mars will begin with our first regolith radical, George Leonard and "Somebody Else In On The Moon." But first, let us take a walk through the real facts of space travel and the possibility of life existing beyond earth's atmosphere.

IS THERE ANY LIFE BEYOND EARTH?

Everyone seems to be in the dark about other life in the universe. People ask many questions about alien life. Are humans the only intelligence in the universe? Are there other beings on other planets like us? Have advanced beings visited us in the past. Are super races from distant worlds watching planet earth? Are there Selenite's hiding under the lunar surface? Are there any simple answers?

Yes, there are simple answers. Some answers are in religious texts, such as the Bible. Other answers are obtainable from scientific research. The first records there are only two forms of life. One is human and material, while the other is nonhuman and supernatural.   There is no indication of physical beings from other planets. Many scientists on the other hand, wish to find evidence for extra-terrestrial beings to avoid the supernatural.

The subject of this book is the universe is sterile of life forms other than what is on planet earth and it is the unique center and origin of life, as we know it. That earth is the center of life makes perfect sense, just as it makes sense that the big bang is the center and single origin of matter.

24

So far, life is exists only on earth, while no evidence of life exists beyond it. Consequently, it is most probable that there is none elsewhere. People are tossing all sorts of inconclusive alien fictions into the arena, but until someone pulls some cosmic alien rabbit out of the outer space hat, the burden of proof will continue to rest upon the proponents of alien presence. So far, every claim to alien life forms has failed the test of science. There has not been one example of absolute conclusive proof presented since ET research began. The air is silent and the table is empty.

Fortunately, for the skeptic, science must argue from silence, and this silence speaks more against than for ET life forms. Moreover, all claims of moon-men, Venusians and Martian invasions have been fictional, and therefore alien artifacts must follow suit.

Let us review only a selection of material as a sample test of the validity of the alien presence theory and the existence of ET artifacts. What will be obtained from this study will be a negative result and this is unacceptable to most people. At least half of humanity hopes that there are alien beings and a smaller percentage actually believe they exist. Some few claim they exist from personal experience! The materials dealt with will be both physical and photographic. When it comes to the photographic evidence for alien artifacts, which is the bulk of this book, the reader will realize that people see what they want to see and not what they should.

Scientific exploration consists of two primary fields of study. The first is a purely materialistic and geological, and consists of the hopes for human colonization and the acquisition of new resources. This is justifiable, since human colonization of other planetary bodies will require other world resources to be successful.

The second is the continued survival of earth itself. The second field is a two-part determination of astrobiologists. It is first to prove the chemical evolution of man and secondly to prove that man can survive in space, therefore consequently proving the higher probability of alien intelligence.

To find any alien life form, other than earth based complex organisms, is to prove the evolutionary theory of life and thus the higher potentiality of ET existence. The hidden agenda behind this ideology is to undermine the supernatural alternative. The agenda

25

of metaphysics, on the other hand, is to disprove materialism in favor of the supernatural origin of biological life. The former is predominately polygenetic by believing in life spontaneously popping up (out of lifeless matter) all over the universe, whereas the latter is usually monogenetic and holds a minority view that life "popped" up only on earth. Both do share and favor some original explosive and creative causal mechanism.

Nevertheless, the primary argument is what comes first as the cause of the other - lifeless matter or intelligence. Therefore, this is ultimately the question: Did life come from lifeless matter, or did all that is physical come from some universal mind? When a person really looks at the data available, it is more "religious" to believe that intelligence came from dead matter, than to believe that matter originated from intelligence.

The continuation of the search for the origin of life evidences a high probability that humankind is alone. The "mind" that is looking toward lifeless matter for the answer is at a loss. The evolutionist cannot have the "mind" as the originator of matter substance (the dirt, rocks, minerals and liquids), since that is where they believe intelligent life came from.

Materialists continue to look for the origins of life in their own way, despite the fact that all efforts so far reveal sterility. The evolutionary biologist finds only lifeless rocks and liquids no matter how deep they dig. Furthermore, for all the species to exist today there must necessarily be a plenitude of fossil records of "smart-rocks."

However, there are no fossil records of educated rocks. An intelligent rock, as the first mutation from dead matter, does not exist. The evolutionary astrobiologists will find only the same lifeless rocks and liquids, no matter how far out in space they probe. Eventually, the astrobiologist must face this fact about space just as the evolutionary biologist must face it with earth. There is no evidence that life (intelligent or otherwise) stems from dead matter. The gap (the missing evolutionary linkage) between lifeless rock and the first living organism is as vast and empty as space itself.

Admittedly, a sterile universe seems more unbelievable than the fancy of one teaming with life, but for the moment, the unbelievable is very true, while the fancy is not. Apparently, what is most probably true is the least desired, whereas the undesirable ends up the very bed we sleep in – all alone in the cosmic sack!

Unfortunately, it seems that silence will continue to beg the question until this circus of pseudoscientists parade a sideshow of alien corpses in the news. How long will that take? Will another thousand exploratory rovers find the illusive microbial PAUL? How many more billions of tax dollars are to be wasted looking for nothing?

People do not seem to tire easily chasing imaginations, fantasies, spooks and space creatures. History shows us many who died looking for the Fountain of Youth, the Pot of Gold over the rainbow, the Serpent Crown, the planet Vulcan and the Holy Grail. It is a fair and just assumption that this will be the case with aliens and consequently, with alien artifacts.

The great silence points more toward a sterile universe, a monogenetic and geocentric origin for man, and a creative intelligence (big banger) for the cause of it all. The bustling alien occupied universe of the UFO nuts has no support. A sterile universe seems to be staring them in the face with no intelligent "cosmic rabbit" in the empty hat of space to demand otherwise. The earth is unique in the universe as the only center and origin of life, and it took divine intelligence to make it happen. So, hats off to you, with no hard feelings, when you hear the bell toll *"And all things came into being by Him, through Him, and for Him"* [1].

Do angels, demons, the Jinn and other supernatural beings exist? Are they extra-dimensional beings or real flesh and blood creatures? There is no physical proof to their existence, though our histories are replete with records of them. Yet, this does not mean they do not exist in some "spiritual" dimension or plain of existence.

Nevertheless, to the metaphysician there is more evidence for ghosts than for all the above. So, what we are talking about in astrobiology? Astrobiology is the study of alien life forms and not angels, demons or Jinn.   Therefore, if aliens exist, they surly claimed and occupied the moon as well as other parts of the solar system or they have not. Moreover, because aliens are unproven to exist on the moon or Mars, some scientists say they reside in the

27

asteroid belt. If they are not in the asteroid belt, they must be elsewhere – because they are out there.

Logic dictates that in traveling, colonization, hunting for resources and establishing outposts, intelligent creatures surly leave conclusive evidence behind as a signature to their existence. Forensic science dictates that every criminal leaves a trace of evidence behind such as a fingerprint, DNA, a fingernail, maybe even bags of litter, trash, garbage, waste, foot tracks or varied other clues. Native Indian "trackers" can follow an animal for miles according to the disturbances made in animal movements.

Now, there is tons of testimony, claims, visions, apparent trace-fossils, supposed implants and even accounts of abductions, but nothing forensically conclusive to prove the existence of alien life forms of any kind.

This great silence is a lesson in adding up negatives, where the result is never positive. Statistical mathematics in exobiology favors more the argument against an alien presence by its stacking negatives and never summing up one positive. The probability of ET life existing slowly decreases to zero the more the negatives pile up. For example, the probability of alien life decreases each time a space mission returns a negative result.

The following evaluation of astrobiology will attempt to disprove the chemical evolution of life by showing the lack of evidence of alien life forms in our solar system. A list of scientific complications for interstellar travel will be enough to show no aliens have visited our planets.

A further examination and comparison of the geology, rocks, minerals and fossils of earth with those of other planets will show the reader that there are no ETs in our solar system. The supposed space bacteria, meteorite fossils, moon lichens, lunar bases, cities, and the many lunar and Martian photographs taken by space probes will demonstrate the same.

This book will conclude, from the complete lack of any biogenetic (biological fossil) evidence from ET materials, that there is practically zero probability in finding life beyond earth. This will establish the falsity of the chemical evolution of man as well as disprove the evolution of life forms in the universe other than that on earth.

This book will conclude with affirming the sterility of the moon and Mars, and that the likelihood of life elsewhere is practically zero, and consequently, that the belief in alien artifacts, is just that, a belief.

There are good arguments for the sterility of the universe. Dr. Dennis Bonnette, author of 'Origin of the Human Species' said it the best. *"Inter-specific evolution [also "naturalistic evolution"] would appear impossible for two reasons: 1.) because the effect (a new and higher form) cannot come from an insufficient cause (the prior and lower form), and 2.) an agent in a given species tends only to the production of effects that remain within that same natural species. For both reasons, non-living agents cannot produce living effects, non-sentient organisms cannot give rise to sentient ones, and non-intellective primates cannot give rise to intellective ones."* [2]

The following negatives indicate that a sterile universe is most probable. Until the advocates of the "alien presence" fiction prove otherwise, the silence of positive evidence will continue to stand against them.

ARGUMENTS AGAINST ALIENS IN OUR SOLAR SYSTEM AND THE IMPRACTICALITY AND IMPOBABILITY OF INTERSTELLAR SPACE TRAVEL

Society is bombarded with what seems to give credibility to alien intelligence. Front covers of popular magazines proclaim all sorts of claims of alien contacts and proofs of ET visitors. Movies like Close Encounters, E.T., PAUL, and others with their dazzling special effects, also fuel the imagination.

The list is endless with stories that further credence to the idea that there are other beings out there. With all the hype, it is tempting to jump on the ET bandwagon no matter who you are. The ET syndrome is a cultural paradigm that seems to span the entire spectrum of society with impunity.

The stars above us are such a beauty that men have fashioned whole mythologies around them. The stars are truly a brilliant sight and now that we have extended our reach to the moon, even landing on the moon, the natural progression is that we might want

29

to travel to them. Such travel is a basic part of countless science fiction stories and films, and many come away with the impression that interstellar travel is an easy task, perhaps just around the corner for man.

Sadly, there are serious problems to consider. Problems that we will see are insurmountable and practically impossible to overcome. Since the laws of physics are the same throughout the universe, the following physical difficulties will show that it is highly unlikely that any other advanced race of beings have mastered space travel. It will also show the high probability that man will not travel deep space or colonize another planet.

A small list of some of the mired problems will suffice in showing the extreme complexity and near impossibility of interstellar space travel for both humans and aliens. The technology necessary to reach another livable planet within an aliens' concept of practicality cannot be too different from ours, according to the laws of physics. Just to consider accepting such a challenge would require the motivation of a doomsday scenario.

The documentary "Evacuate Earth" presents just such a scenario and just how almost impossible it would be for man to escape earth within a 75 years period from an approaching global killer neutron star. Just the idea of building a prefect (100 year plus) generation space ship (they call it "the Horizon Project") to reach another planet is, needless to say, ridiculous and totally impractical at this time.

When the details are really considered, beyond the lame simplicity of the docudrama, the whole idea of an interstellar generation spaceship should rather be considered a suicide ship. Just getting to another planet is "Treky" to begin with, not to mention determining the perfect planet to travel to. Moreover, creating the perfect ecosystem on the new planet (terraforming) to accommodate our species is stretching time twice that spent getting there.

Even though the above docudrama leaves us with some small bit of hope, and a false hope at the moment, what it forgets to mention, after solving all the trivial problems of building the ship, supplying it with all our provisions, is deep space protection from radiation, space cancer, artificial gravity deformation of human physics and interstellar disasters as we shall soon demonstrate.

The show, by the way, ends with the evolutionary rhetoric of the survival of the fittest "necessarily" succeeding in reaching the new livable planet in light of and in expectation of expanding out into the Universe forever – for they cannot see that there is an end to humanity as we know it.

If beings have come to our solar system from some distant star system, which does n't seem to "sound" ridiculous, it would define them as so advanced, that if they were observing us, we would not amount to much more than a curiosity, if not a hindrance to their colonization. Considering us even closely "intelligent" or at least valuable would be straining their better judgment.

Furthermore, if life is universal and abundant, and livable planets are at a premium, we would be expendable with no one to protect us. To such advanced races, we would be no more worthy to them than insects are to us. If they could even get here from such vast distances, believe me, we would be very expendable if their lives depended on it.

DISCOURAGING DISTANCES

Is it possible to build an engine practical and safe enough to get humans to other star systems? The obstacles in traveling the enormous distances in space are huge, so the prospect that space aliens are visiting earth is highly unlikely. Even if one assumed the existence of life somewhere else in the universe, a visit by ETs to earth, such as is claimed in UFO reports, seems impracticable, if not impossible. The distances and consequently, the likely travel times are unimaginably vast.

For instance: The first on the list of rest stops is the closest star system Proxima Centauri, which is 40.7 million, million kilometers (c. 25 million, million miles) away. This is 4.2 light years away. This measurement designates the time it takes light to travel one year, which is 186,000 miles per second. At the incredible speed of one-tenth of the speed of light, the trip would still take 43 years. Proxima Centauri is the third star in the Alpha Centauri star system, also known as Alpha Centauri C. Lucky for us, Alpha Centauri B has one confirmed planet. This was the hope

31

behind the science fiction TV series, "Lost in Space." Good luck to whatever Robinson family-II foolishly sets off to that destination! [3] There is no assurance that the planet is habitable.   The distances to the other nearest star systems, especially those with livable planets are vast. In fact, the distances are out of this world and beyond practicality.

Other planetary star systems also offer very little hope, if any at all, of the likelihood of man colonizing outside this star system. There are no livable planets close by and what potentially livable ones there might be are too far away – and this is an understatement.

(2.) If Alpha Centauri is not good enough, the next closest star is Barnard's Star. It is 5.9 light years away. It is a faint red dwarf star, discovered in 1916 by E. E. Barnard. Recent efforts to discover planets around Barnard's Star have failed. Obviously, this is no good either. What about colonizing the next star system?

(3.) Wolf 359 is 7.7 Light years away. Wolf 359 is also a red dwarf. It is so small that if it were to replace our sun, an observer on Earth would need a telescope to see it clearly. The star is too small for any habitable planets to exist. Let us keep on trucking to the next hopeful star base.

(4.) Lalande 21185. Distance: 8.26 LY. While it is the fifth closest star to our own sun, it too has no planets. We could plant a space base and refueling station here as a stepping stone to the next planetary system. There are four more star systems with no planets:

(5.) Luyten 726-8A and B., 8.73 LY.

(6.) Sirius A and B. No known planets. 8.6LY.

(7.) Ross 154. 9.693LY, and

(8.) Ross 248, is a whopping 10.32 light years away.

(9.) Finally, there is the first star system with a known planet, Epsilon Eridani. Hot dog! Maybe we found a new home? Let's check it out. Epsilon Eridani. Distance: 10.5 LY. Eridani (tenth closest star to Earth) is the next closest star known to have a planet, Epsilon Eridani B. It is the third closest star that is viewable without a telescope. It would take about 100+ years to get there at 1/10th speed of light, considering no stops along the way or accidents. Nevertheless, if this planet offers no hope, we must travel further and further, and further, and keep on going.

(10.) Tau Ceti is next on the list, which is 11.8 light years is a single G8 star similar to the Sun. High probability of possessing a solar system type planetary system: Current evidence shows 5 planets with potentially two in the habitable zone. Let us round this off to 12 Light years away. This is 4 light years (25 Trillion miles) times 3 = 75 Trillion miles away.

(11.) Gliese 581, 20.3 LY away. Multiple planet system. The unconfirmed exoplanet Gliese 581g and the confirmed exoplanet Gliese 581d are in the star's habitable zone. This one is 5 times 25 trillion miles away = 125 trillion miles.

(12.) Vega is 25.0 LY away. It has at least one planet and is of a suitable age to have potentially evolved primitive life – if you happen to believe in the fiction of evolution. This one is about 150 trillion miles away. If it takes 44 years to reach the nearest star, it will take 6 x 44 years or about 265 years to reach this one. We do not want to calculate the fuel, food and multi-tons of other supplies needed to get there. Now, what if this is a dead end planet? We must keep traveling, right?

FICTIONAL FEASABILITIES

Space travel is amazingly difficult if not a fantasy in the minds of fiction writers. Since humans have traveled to the moon, and have sent probes on short trips within our Solar System, we tend to underestimate how big and how hostile interstellar and intergalactic space is.   Though there may be physics we have not yet discovered that allows us to travel safely among stars, as of now we do not see any evidence that such physics exists.

The vast distances between planetary systems screams against the survival of any physical creature traveling outside their planetary system. A visit or colonization to our solar system by ETs is impractical and most improbable, no matter if possible. Just eliminate what is impossible and whatever is left is most probably the answer. It's too dammed far and costs too much and is purposeless.

As we mentioned, the distances and travel times are unimaginably vast. Even the closest star system Proxima Centauri is too far away. Mathematical calculations show Proxima Centauri to be 25.12 TRILLION miles from our sun. To understand this kind of distance let us use some comparisons.

If you could shrink the galaxy down to the point that our Sun is the size of a golf-ball, then Proxima Centauri would be a golf-ball about 30 miles away. How far is that really? This gives us a "visual" idea.

For example, the sun is 93 million miles away from earth, so let us compare that to Proxima Centauri. To get an idea how far this is, first picture the measurement of 93 million miles as a one-foot ruler. How many one-foot rulers would you need to lay in series to equal that distance? First, figure how many "sun to earth" distances there are by dividing 25.12 trillion by 93 million. How many rulers are you going to need? The answer is 270,108 rulers! That is actually 93 million miles TIMES 270,108 thousand! Just take the 108 times 93 million and you have about 10 billion miles. A trillion is 1000 billion. We are talking about 27 times this distance.

Well, we can all agree that it really was a great accomplishment for Neil Armstrong to set foot on the moon, and maybe a bigger step for mankind, but there will be Mules populating the moon first before humans ever step foot in another solar system!

Despite this impossible hump, and the sterility of Mules, many Mule-headed dreamers believe humans can do it if we have the right propulsion system. Yes, this sounds true, but it is more believable that the ass spoke to Balaam than to swallow that whale of a promise.

When busying ourselves in our man-cave watching this "Lost in Space" scenario to Alpha Centauri, let us not forget to bring plenty of popcorn. And plenty we will need. Just how much popcorn do we need to see this trip from start to finish? First let us calculate the time it will take us from the distance there and with the technology we have.

Calculating the time, size of family and how much each person must eat to stay alive during the trip, we should come up with the proper amount. But, would this be reasonable? The imagination is a wonderful gift from God for creative thinking and entertainment, but that is all it is good for, when planning a trip like this.

You are thinking this is silly reasoning, and that it is possible to join the Robinson family. Let us see which is silly. Imagine this: The Apollo flights took three days to get to the moon. At the same speed, one would need 870,000 years to get to our nearest star. A Mule would just as well get you there! Munchhausen used a string of geese, and he was not that far removed from telling us the truth.

Nevertheless, humans are inventive and he could increase the Apollo speed to one-tenth the speed of light. How long would it take humans to get there? At the incredible speed of one-tenth of the speed of light, the trip would take about 43+ years!

OK, so my son would live to enjoy landing on the new planet. I would have to leave here at the age of at least 25, because when we add 43 years to that, we are talking about being 68 years old when we get there. Of course, I would have to refrain from procreating until 20 years has lapsed, least I over populate the spacecraft, etc.

To continue our calculations, what is one-tenth of the speed of light in miles per hour? In imperial units, the speed of light is approximately 186,282 miles per second. This is 11,176,920 miles per minute, times 60 minutes in an hour = 670,615,200 miles per hour. My God! That is FAST! I just hope we do not hit a chuck hole along the way.

OK, so our flimsy Apollo spacecraft has to travel at 670 million miles per hour to get there within a reasonable length of time - (cough!). Let us just say that our spacecraft can travel at 93 million miles per hour or one trip to the sun in an hour. How many "earth to sun" distance "hours" does it take to get to this star? The answer is 270,108 hours. Divided by 24 hours this comes to 11254.5 days, which is 30.83 years.

So, we have calculated the distances, the time and speeds, and now we can determine the amount of popcorn to take along.

SPACE TRAVEL IS POPPYCOCK!

How much popcorn must we take to make the above trip, if only six people go, and each two people eat one pound of popcorn every two days, until they get there? 11.77 tons! Now remember, this is just popcorn. We have not even started adding up all the other foods and supplies we will need. We have also not considered that it will take more than six people to get there. Try calculating the amount of supplies for 100-200 and maybe more than 300 people.

Is it possible for man to set foot in another star system taking along that much popcorn, and of course arriving with his full faculties? It is pure poppycock to think so! Nevertheless, anything is possible on paper, in a movie and within the imagination of scientific fiction writers. Who knows what missing element can "make it so," as Captain Jean-Luc Picard said. I believe he is actually replying to the question of butter.

This brings up the vital question of how much butter we must take along with the popcorn, not to mention the cola to go with the sideshow. Of course, the purpose of the popcorn is to feed the travelers while they are entertained with movies. So, how many (new) movies would be needed for entertainment, if one movie is shown every two days? 7,847 movies and one short 30 minute commercial, of course showing us popcorn.

NASA and other scientists have all sorts of wild new fantastic "theoretical" propulsion systems on the drawing board. That is where they will stay too, on a drawing board. Nevertheless, just to be sure, why not rather ask the real first man on the Moon, the Baron, Hieronymus Carl Friedrich von Münchausen.

Seriously though, let us consult the best authorities at Acme Scientific Co., and see what engines are available. Once we exit science fiction theatricals, physics tells us what we can and cannot do, as well as what aliens may do and not do.

SPACECRAFT AND PROPULSION SYSTEMS

The Jet Propulsion Laboratory in California studied the feasibility of an ion-thruster vehicle as a forerunner of a star-ship. Called the TAU (Thousand Astronomical Unit) mission, its aim was to carry instruments to the edges of interstellar space. In calculating its capacity, they concluded that if traveling at 105 kilometers per second, it would take more than 10,000 years to reach our nearest star. This is one down! But, let's see what other options we have.

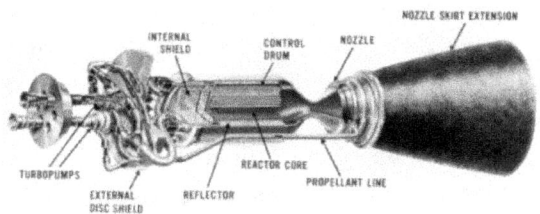

There are other kinds of space propulsion systems on the drawing board for traveling to another star system. Spacecraft for interstellar travel needs an efficient propulsion method. No such spacecraft exists yet, but there are many designs. Since interstellar distances are astronomical, a tremendous velocity is required to get a spacecraft to its destination in a reasonable amount of time. Acquiring such a velocity on launch and getting rid of it on arrival will be a formidable challenge for spacecraft designers.

There are many proposed designs for getting humans to another star system. Let's analyze each to see which one is most efficient and affordable.

Nuclear Rocket Engine:

One particular design is the Nuclear Rocket Engine. However, is this engine sufficient to catch up with the Robinson family? The high exhaust velocities necessary to reach high speeds, especially for interstellar distances, makes important engineering considerations with rocket concepts of all types very difficult.

Regarding the acceleration time issue, accelerating at 1 gravity (1g) for one year, a ship would reach just somewhat less than the speed of light. Thus, even a rocket with a theoretical top speed of 0.5c, but limited to a peak acceleration of 0.001g, would take ~500 years to reach maximum velocity. [4] The price listed in the Acme Scientific Co., Catalogue #1 is $600,000,000. Add $600,000,000 more for an extended warranty. This does not include the shipping cost of sending it back 27 trillion miles. Obviously, this one is too slow and too costly.

Thermal Fission Rocket Engines:

Fission-based thermal rocket concepts incorporate heating an exhaust gas such as hydrogen like the NERVA nuclear rockets studied during the 1960s, while potentially able to achieve high accelerations, have fairly low exhaust velocities (ultimately perhaps up to a few tens of times greater than the best chemical rockets) and are thus unpromising for missions over interstellar distances. [5] Acme Scientific Co., in Catalogue #2, discontinued this device. Obviously, not very promising.

Fission-electric Engine:

Nuclear-electric or plasma engines, powered by fission reactors, have the potential to reach speeds much greater than chemically powered vehicles or nuclear-thermal rockets. With fission, the energy output is approximately 0.1% of the total mass-energy of the reactor fuel and limits the effective exhaust velocity to about 5% of the velocity of light. This means that achieving start-stop interstellar trip times of less than a human lifetime require mass-ratios of between 1,000 and 1,000,000, even for the near stars. This is achievable by multi-staged vehicles on a vast scale.

The disadvantage is that it is too massive and too expensive. [6] Again, the Acme Scientific Co., Catalogue #3, priced this at $600,000,000, but does not expect any buyers. This one costs too much and is now discontinued.

Nuclear Pulse Propulsion

NPP systems contain the prospect of very high specific impulse and high speed, and therefore of reaching the nearest star in decades rather than centuries. **Project Orion** was a study of a spacecraft intended to be directly propelled by a series of explosions of atomic bombs behind the craft (nuclear pulse propulsion).

The Orion concept offered high thrust and high specific impulse, or propellant efficiency, at the same time. The unprecedented extreme power requirements for doing so would be met by nuclear explosions, of such power relative to the vehicle's mass as to be survived only by using external detonations without attempting to contain them in internal structures.

From 1957 until 1964 this information was used to design a spacecraft propulsion system called "Orion", in which nuclear explosives would be thrown behind a pusher-plate mounted on the bottom of a spacecraft and exploded. The shock wave and radiation from the detonation would impact against the underside of the pusher plate, giving it a powerful "kick". The pusher plate would be mounted on large two-stage shock absorbers that would smoothly transmit acceleration to the rest of the spacecraft.

However, interstellar travel would only be possible using advanced derivatives of present designs with cruising speeds of 8%–10% of the speed of light. [7] Present designs have exhaust velocities of 20–30 km/s, far too low to achieve reasonable interstellar cruising speeds. [8] A 1968 study estimated that it would take on the order of 1330 years for the energy-limited heat sink Orion design to reach Alpha Centauri.

The "Momentum Limited" Orion design was estimated at 133 years to reach the nearest star. (8a) Modern proposals utilizing Z-pinch fusion schemes are also under development, though again, the technology may be more appropriate for outer Solar System exploration than true interstellar flight.

As with rocket concepts generally, the high exhaust velocity vs. high acceleration issue noted above relative to specific exhaust power applies, and is to likely limit realistic interstellar applications. This one is too slow with present technology.

Fusion Rockets:

Fusion rocket star-ships, powered by nuclear fusion reactions, should conceivably be able to reach speeds of the order of 10% of that of light, based on energy considerations alone. In theory, a large number of stages could push a vehicle arbitrarily close to the speed of light. The maximum exhaust velocities potentially energetically available are correspondingly higher than for fission, typically 4-10% of c.

However, the most easily achievable fusion reactions have a significantly high energy loss. Thus, while these concepts seem to offer the best prospects, they still involve massive technological and engineering difficulties, which may turn out to be intractable

39

for decades or centuries. This one is too difficult and has too many problems to resolve.

Ion Drives:

An ion thruster is a form of electric propulsion that creates thrust by accelerating beams of ions to create thrust. For Earth launch, the engine would have to be supplied with one to several gigawatts of power, equivalent to a major metropolitan generating station. Nevertheless, outside the earths' atmosphere, ion engines are practical, at least within proximity to our solar system.

A major drawback with ion propulsion systems is the lifetime of electrostatic ion thrusters. The engine becomes dysfunctional when holein the accelerator grid become so large that the ion extraction is largely affected. Grid erosion is unavoidable and is the major lifetime-limiting factor. Hall thrusters suffer from very strong erosion of the ceramic discharge chamber.

A good example of ion propulsion for inner solar system use is the Dawn spacecraft, launched on 27 September 2007, to explore the asteroid Vesta and the dwarf planet Ceres. Dawn's ion drive is capable of accelerating from 0 to 60 mph (97 km/h) in 4 days, firing continuously. [9] Apparently, this model is unsustainable and cannot be used for long distance.

Antimatter Rockets:

An antimatter rocket would have a far higher energy density and specific impulse than any other proposed class of rocket.
 Energy resources and efficient production methods of antimatter make these rockets impractical and thus nonexistent. With the proper amount of antimatter, it would be theoretically possible to reach speeds approaching that of light.

Two further problems need a solution. First, in the annihilation of antimatter, much of the energy is lost. Even so, the energy available for propulsion would probably be substantially higher than nuclear fusion, the next-best rival candidate. Second, heat transfer seems likely to deposit enormous wasted energy into the ship - gamma rays. Third, biological shielding is necessary to protect the passengers. Energy would inevitably heat the vehicle, and may thereby prove limiting. This requires consideration for serious proposals if useful acceleration is to be achieved.

The energy involved is also very large. The other problem is, the antimatter must be electromagnetically kept suspended and from touching matter – the inside walls of chambers, or it instantly explodes. If the magnetic suspension chamber fails and the two combine, the ship would be destroyed. So, for 100 years plus, not one failure must be allowed to happen. Anti-matter engines are at present are practically impossible and highly dangerous.

Magnetic Monopole Rockets:

If some of the Grand unification models are correct, we can construct a photonic engine that uses no antimatter, but uses a magnetic monopole, which hypothetically can catalyze decay of a proton to a positron. As a result, a hydrogen atom turns into 4 photons and only the problem of a mirror remains unresolved.

Nevertheless, most of the modern Grand unification theories predict no magnetic monopoles, which casts doubt on this attractive idea. [10] As far as anyone knows this is not possible.

Interstellar Ramjets are Inefficient.

In 1960, Robert W. Bussard proposed the Bussard ramjet, a fusion rocket in which a huge scoop would collect the diffuse hydrogen in interstellar space, "burn" it on the fly using a proton–proton fusion reaction, and expel it out of the back. Later calculations with more accurate estimates suggested that the thrust generated would be less than the drag caused by any conceivable scoop design. [11] Totally inefficient and useless for star travel. But, Acme Scientific Co., has the theoretical plans. PDF download if you can locate them.

Beamed Propulsion:

This diagram illustrates Robert L. Forward's scheme for slowing down an interstellar light-sail at the destination star system. A light sail or magnetic sail powered by a massive laser or particle accelerator in the home star system could potentially reach even greater speeds than rocket- or pulse propulsion methods, because it would not need to carry its own reaction mass and therefore would only need to accelerate the craft's payload.

Robert L. Forward proposed a means for decelerating an interstellar light sail in the destination star system without requiring a laser array to be present in that system. In this scheme, a smaller secondary sail is deployed to the rear of the spacecraft, while the large primary sail is detached from the craft to keep moving forward on its own. Light is reflected from the large primary sail to the secondary sail, which is used to decelerate the secondary sail and the spacecraft payload. [12]

A magnetic sail could also decelerate at its destination without depending on carried fuel or a driving beam in the destination system, by interacting with the plasma found in the solar wind of the destination star and the interstellar medium. Unlike Forward's light sail, this would not require the action of the particle beam used for launching the craft. Alternatively, a magnetic sail is pushed by a particle beam or a plasma beam to reach high velocity, as proposed by Landis and Winglee.

Beamed propulsion seems to be the best interstellar travel technique presently available, since it uses known physics and known technology developed for other purposes and would be considerably cheaper than nuclear pulse propulsion. Possible, but highly improbable.

Anti-gravity Propulsion:

So far no devices have surfaced and most probably never will. At present they are non-existent.

ENERGY / FUEL REQUIREMENTS

A traditional system involves burning fuel or reaction mass, but to reach another star, impractically vast quantities are required. One solution is to pick up fuel along the way. In the space between stars, there are not convenient asteroids and planets to land on and mine for fuel.

Luckily, space is not quite a vacuum, and there exist tiny atoms scattered far apart, mostly hydrogen. Going at a fast speed, these atoms could be gathered and used as fuel in an efficient reaction such as fusion (presuming we achieve fusion someday). To collect them, a huge scoop is necessary, and conservative calculations put it at least 2000 square km in area, which would cripple the ship with its drag and limit the speed to being slower than the space shuttle.

This system is horrendously inefficient and not viable considering that our sun is placed in a sparse region of space, providing a poor fuel source. The enormous energy necessary to reach Proxima Centauri would be roughly equivalent to the electricity output of the world's largest hydroelectric power station for four days.

One honest scientist warns, *"For a manned spacecraft weighing many tonnes, the energy requirements would exceed the world's annual electricity consumption. For the city-sized spacecraft in [the movie] Independence Day, the energy requirements would be even more staggering. In addition, when the spacecraft slowed again, it would need to use up almost this amount of energy in braking. If the spacecraft had to accelerate to c/10, slow down and start up many times, it is hard to imagine how enough fuel could be carried."* [13]

It seems that if it must take the total energy requirements of planet earth to get there, it would be more feasible to jettison the entire planet earth in that direction somehow.

The biggest problem contributing to interstellar travel is not so much the propulsion system, but the energy needed to obtain a reasonable travel time. The energy needed to accelerate the craft to reasonable travel speed is equal to the energy needed to decelerate. This means the craft has to have at least twice the fuel to reach its (one-way) destination.

The velocity for a manned round trip of a few decades to even the nearest star is thousands of times greater than present space vehicles. Millions of times as much energy is required for star travel. Accelerating one ton to one-tenth of the speed of light requires at least 125 billion kWh. The spacecraft must carry the necessary energy, as solar panels do not work efficiently in deep space. [14]

There is some belief that the magnitude of this energy may make interstellar travel impossible. In 2008, the Joint Propulsion Conference concluded that it was improbable that humans would ever explore beyond the Solar System. Brice N. Cassenti, of Rensselaer Polytechnic Institute, stated, *"At least 100 times the total energy output of the entire world [in a given year] would be required for the voyage* (to Alpha Centauri)." [15]

DANGERS OF LESS THAN LIGHT SPEED

Big bangs come in small packages! Micrometeorites can be extremely dangerous if not deadly. The kinetic energy of a particle with a mass of a tenth of a gram traveling at a tenth of the speed of light, calculated from the spacecraft's reference frame, is: $1/2mv2$, which = $1/2$ x $10–4$ kg x $(3$ x 107 m/s$)2$, which is = 4.5 x 1010 J. In a chemistry lexicon (Roempp Lexikon), the combustion energy of TNT is given as: 4,520 kJ/kg = 4.52 x 109 J/tonne. Thus, 4.5 x 1010 J is equivalent to: $(4.5$ x $1010)/$ $(4.52$ x $109)$, which is = 9.95 tons of TNT. Therefore, the impact energy of one of those 0.1 g objects would be the equivalent of an explosion of about 10 tons of TNT. A one gram particle would be 100 tons of TNT.

What is a 10 ton of explosion like? If, you can imagine what it's like when dropping a big house brick on a tiny ant. The effect is the same. Another example is it is enough to obliterate a battleship. Well, a 10 ton TNT explosion is bad enough, but what about the following problems?

We know now what a one-tenth of a gram size particle will do to our spaceship. Let us add more "bang" for the buck to our argument to illustrate the absurdity of interstellar space travel. The question is, what possibility is there of survival with a larger size particle slamming into our craft.

Let us calculate: A one gram particle hit: .1g x 10 tons TNT x 10 = 100 tones TNT. A one ounce meteor hit: .1g x 10 tones TNT x 10 = 100 tons TNT x 28g = 2,800 tons TNT. This is equal to a small nuclear bomb.

Now, average ONE rock (about one ounce) hitting the craft every one million mile. We should figure one in a million miles could be a fair appraisal of the frequency of particles between 1/10th of a gram to a one ounce size hitting the "front" of the craft, but let us say, that the worst case is, 1 one-ounce rock hitting every one million miles.

Remember, the craft is traveling at 1/10th the speed of light, and that is 67 million miles per hour. This equals 67 x 2800 tons of TNT hitting our craft per hour = death! Well, at least it is 187,600 tons of TNT. This is a lot of restriction as well in traveling forwards. This would have to be counteracted by equaling if not superseding this amount in propelling the craft. So, one must figure not only the base fuel, but the extra for compensation.

We are being hit 67 times per hour with a 2.8 kiloton nuclear bomb! Check out the internet for video examples of 10, 100, 300 and 1000 kt bombs. Now, calculate 67 times per hour, times 270 thousand hours or 43 years. Notice the energy and danger facing us: 187,600 tons TNT x 270 thousand hours = 50,652,000,000 tons of TNT! 50 billion tons of TNT? Come on people, "alien" or no, we cannot go.

Thus, we have only two choices. We take our time and die or we hurry up, and suffer the above, and die. Even the least dangerous brings us into contact with micrometeorites, which are very deadly traveling at extreme speeds.

We may argue that "most" particles will be made of magnetic materials, and deflectable with a high-energy electromagnetic field. Nevertheless, this does not protect us from those made of non-magnetic materials. The craft cannot dodge these at the speeds talked about. If we cannot deflect them, then what can we do to protect ourselves? We might travel in a plastic water-lined balloon and let them fly right through our spacecraft. The best plan is to let it pass through and hope no one is standing in the way!

The big question is, how long before the above frequency of hits tares our cosmic swimming pool apart. Speed therefore is not the answer. Going slower must be safer. Remember, every few million miles the craft slams into a small nuke.

However, the die-hard space traveler will say, the above is not fair because it is not true that every million miles the craft is hit. This is a good argument. Nevertheless, one hit is potentially deadly. There is no assurance that the one rock will not ever be the fatal rock. It is false hope that said humans can make it there without facing a fatality.

To travel to another star system humans need 100% safety. On earth, all we need is a large percentage, because we have other humans to call to the rescue. In the depths of space, there is no one to rescue a fatality. Besides, the only 100% guarantees that exists are death and taxes, and the latter is questionable with the right attorney.

What does Murphy's Law tell us? The probability of a negative happening depends upon the percentage of the possibility. Is it possible for one huge doomsday meteor to destroy an interstellar spacecraft? Is there a chance that you might win the lotto despite the odds against you? Playing the odds with the lotto is one thing; playing with the odds of making it alive to another star system is another. You have a better chance of winning the lotto than making to Alpha Centauri.

DANGERS OF LIGHT SPEED

Many stories include zany explanations of how faster-than-light travel is possible. The reality is that physics prevents this. There are no cheats. Even close-to-light travel runs into all sorts of interesting relativistic problems involving mass and energy, as well as increasing radiation problems. Staying within the realm of matter, traveling at almost light speed intensifies whatever problem one would have at lower speeds. It turns atomic particles in to mini atomic bombs. Light speeds increase the density of particle bombardment and does subject spacecraft occupants and equipment to extreme levels of radiation.

Dr. Oleg Semyonov said, *"When a ship accelerates to a relativistic velocity above 0.3c, interstellar gas becomes a flow of relativistic nucleons, hard radiation bombarding the starship, In addition, interstellar space contains high-energy cosmic rays and dust, all of which can present a huge problem if no proper protection is implemented... a radiation dose obtained in a non-relativistic space module...would be, approximately, 70 rems/year...(the safety level for a person is between 5 - 10 rems/year). The dose will, most likely, increase when accelerating to relativistic velocities...utilize extreme relativistic velocities...would present significant radiation hazards onboard."* [16]

DANGERS OF WARP SPEED

Warp speed is a sci-fi term for faster than light speed travel. Star Trek fans, prepare to be disappointed. Kirk, Spock and the rest of the crew would die within a second of the USS Enterprise approaching the speed of light. The problem lies with Einstein's special theory of relativity. It transforms the thin wisp of hydrogen gas that permeates interstellar space into an intense radiation beam that would kill humans within seconds and destroy the spacecraft's electronic instruments.

Interstellar space is not empty. For every cubic centimeter, there are fewer than two hydrogen atoms, compared with 30 billion x billion atoms of air here on Earth. According to William Edelstein of the Johns Hopkins University School of Medicine in Baltimore, Maryland, that sparse interstellar gas should worry the crew of a spaceship traveling close to the speed of light even more than Romulans de-cloaking off the starboard bow.

Special relativity describes the distortion of space and time for observers traveling at different speeds. For the crew of a spacecraft ramping up to light speed, interstellar space would appear highly compressed, thereby increasing the number of hydrogen atoms hitting the craft.

LIGHT SPEED / WARP SPEED DEATH RAYS

Another insurmountable problem is that the atoms' kinetic energy also increases. For a crew to make the 50,000-light-year journey to the center of the Milky Way within 10 years, they would have to travel at 99.999998 per cent the speed of light. At these speeds, hydrogen atoms would seem to reach a staggering 7 teraelectron volts – the same energy that protons will eventually reach in the Large Hadron Collider when it runs at full throttle. *"For the crew, it would be like standing in front of the LHC beam,"* said Edelstein.

The spacecraft's hull would provide little protection. Edelstein calculates that a 10-centimeter-thick layer of aluminum would absorb less than 1 per cent of the energy. Because hydrogen atoms have a proton for a nucleus, this leaves the crew exposed to dangerous ionizing radiation that breaks chemical bonds and damages DNA. *"Hydrogen atoms are unavoidable space mines,"* said Edelstein.

The fatal dose of radiation for a human is 6 sieverts. Edelstein's calculations show that the crew would receive a radiation dose of more than 10,000 sieverts within a second. Intense radiation would also weaken the structure of the spacecraft and damage its electronic instruments. Seriously, the potential damage is astronomical. It is equal to stepping into a nuclear furnace.

Edelstein speculates this might be the one reason why aliens have not paid us a visit. Even if Mr. E.T. has mastered building a rocket that can travel at the speed of light, he may be lying dead inside a weakened craft whose navigation systems have short-circuited.   Nevertheless, science fiction said there is nothing impossible for aliens. [17] Maybe their spacecraft runs off antigravity and is controlled by photon based quantum computers? (17a) maybe the aliens have mastered radiation shielding and solved food storage problems? Maybe if my grandmother had wheels, she would be a wagon too? – Engineer, Scotty.

ALTERNATIVES TO CONVENTIONAL PROPULSION

Alternatives that increase the theoretical possibility of such travel are also not practical. They are too dangerous or are technologically impossible. For instance, science fiction is offering ideas.

WORM HOLE SHORT CUTS

Our only possibility is to use wormhole portals. Such a wormhole would have to be carefully controlled, which is beyond our present capabilities, and we would have to somehow manage to create a twin wormhole far off at our desired destination, which might require someone else at the other end. Needing someone else to be there beforehand is not feasible for the first interstellar flight. Worse, the physical effects of traveling through a permanent or semi-permanent wormhole would warp and destroy any matter. You would arrive at your destination in the form of plasma.

TELEPORTATION – BEAM ME THERE, SCOTTY!

Classic teleportation involves a person activating a device and vanishing only to reappear simultaneously at some destination. There are a few movies that depict this fantasy: "They Live" and "Star Trek" are just two examples. This is not as straight forward as it first appears. First, the atomic structure of the person in the teleport machine is unassembled. Secondly, these atoms are teleported to their destination and reassembled.

The reassembly, however, requires a machine to be at the destination location and there are physical laws that do not permit the manipulation of matter at such a fine level, and over such vast distances. Teleportation can only be to places already visited. The reassembly is currently beyond our ability, but might be possible.

The atoms would still have to travel to another star, which might be faster than traveling as a body, but would still take years at least. The closest star to the sun is four light years away, so anything sent would take longer than four years to get there.

Alternatively, the reassembling machine could have a store of atoms from which to assemble the person, but this is in essence creating a copy and destroying the original – there is a movie that depicts this load of baloney. Many people would not be comfortable with this. There is no assurance that the teleport machine would carry the soul with it. No human is assured the soul is purely a derivative of biological processes. It could be that biology is the product of "soul" or "spirit." Thus, there is no guarantee that the teleport machine is not an incinerator.

Another possibility is the "Fantastic Voyage" concept applied to outer space travel where the space ship and occupants are reduced in size to the molecular level and inserted into a laser beam and transported by the laser beam to the destination – this would be only possible after real humans landed there first and set up a receiver unit to re-enlarge the travelers. Pfff! We need to grab another drink and plug in the next episode of the Outer Limits.

GENERATION SHIPS

If faster-than-light travel is impossible or at least impractical, we might look towards generation ships. Driving like Hell usually leads to Hell. So, wisdom said, "Take your time." Even though our nearest star takes light only four years to reach, heavy objects

49

would take much longer. Most stars would take hundreds of years to reach. Generation ships are an alternative designed for a population to live in for generations, until the descendants reach the destination.

There are several problems with a generation ship. The descendants might forget the original purpose of the mission as it fades into legend over the centuries. A cleverly designed computer system might be able to educate people born on the ship to avoid this, but it still becomes increasingly difficult to predict what might occur as the generations pass.

If there is a problem with the ship, the population, which has descended into savagery, will be helpless. This is very possible since computers can break down and people might forget how to fix them. Take vacuum tube technology for instance. With the passing away of the older generation it is becoming almost a dead science. As to nano-tube size circuitry and molecule size microchips?

TEST-TUBE "EGG" SHIPS

To remove as much uncertainty as possible in generation ships, designers have created the concept of "egg-ships." These egg carrying cosmic Noah's Arks would carry frozen human eggs (let us not forget the "sperm-bank"), which would be nurtured by carefully designed machines, acting as wombs, parents and educators. The human eggs can be grown for years, before the distant star or planet is reached, and computers could teach them about their mission, how to survive, and what to do, as well as their history.

Designing care-giving machines that would not emotionally stunt the new humans is well beyond our technology, but perhaps not impossible in the future. Again, there is another science fiction movie that depicts the computer defaulting, turning the mission into a horror movie.

However, like the generation ship, an egg ship does not help the individual, who wants to travel to the stars himself or herself. Waiting for artificially raised humans to live the dream of reaching the stars long after the parents have died is unacceptable to many people.

DEEP SPACE POPSICLES

The above pseudo-scientific solutions are for now impossible. However, there is space stasis or suspended animation, better known as cryogenics. In Laymen's terms, and fitting well with the popcorn and hot butter mentioned above, there is the freezing of humans into Popsicles. What we would have as a spacecraft is a flying space freezer. When longevity and using another generation are not possible, many films and stories depict humans kept in suspended animation to explain long trips.

There are many good alien movies where people are shown waking up to horrors beyond imagination. People would not be able to age in such a state, or would age very slowly, and it would be much like hibernation. Unfortunately, "telomeres" again present a problem. Our bodies always contain a small number of radioactive elements. These emit tiny amounts of radiation, which are harmless, because our cells continually replace damaged ones. If a person does not age in stasis, then their telomeres cannot be shortening and so their cells cannot be dividing.

It follows that any radioactive elements would cause permanent damage to the body, and if given enough time, could result in death. Even slow aging would not keep up with radioactive damage over a long time. We need our cells to divide at a normal rate.

LONGEVITY – SUPER LIFE EXTENSION

An alternative solution is genetically enhance people to live for hundreds or thousands of years so that they could make the journey in their lifetime, assuming the current problems of living in space were solved. Longevity and immortality are both subjects of much scientific research, but their biggest obstacle is *telomeres.*

Telomeres are sections on the ends of your DNA, which are cut slightly shorter each time your cells divide. Eventually the telomeres' lengths shorten and your cells begin damaging their own vital DNA as they divide. This means that our own DNA limits the number of cell divisions we can make. Cells divide to replace old or damaged cells, such as when you brush your skin on something or the constant replacing of your stomach lining cells due to the high acidity in the stomach [18].   The answer seems to be in

51

keeping telomeres long, but generally, the only adult cells which can do this are cancerous.

DON'T BE COOL BE "Q"
IMMORTALITY AND INCORRUPTABLILITY

Traveling between stars will take the gift of immortality and incorruptibility in the biblical resurrection to accomplish. Yet, even with both of these gifts, there might still be problems getting to another star. Even though we cannot die or go insane, we are still corporeal beings, not magical beings or omnipotent masters over matter like the Krell, in Forbidden Planet. But, then again, look what they did to themselves! We could not snap our fingers and transmute matter nor transport ourselves over long distances without propulsion systems. We might be immortal and incorruptible, but we not the "Q" as depicted in the Star Trek series.

Both of the above must be so, to attempt interstellar space flight, for each is different in function. It is true that being immortal is not ever dying, but that does not mean you are immune to suffering corruption, misery and pain.

Alternately, to be incorruptible does not mean you are immune to death and dying. One refers to the physical, while the other refers to the mind or spirit. Without incorruptibility, sin can run just as rampant in immortals with mental illness chasing behind it as it does with a mortal. Check out the insanity of Mr. Q.

On the other hand, without immortality, one could live a perfectly happy, joyful, and painless, but temporary life. Immortality gives you all the time in creation to get there, but does not insure you will appreciate it when you do - it does not filter out evil. It gives you the quantitative value, but not the qualitative aspect+ to get there.

Qualitatively speaking for the incorruptible mortal, he would know better than to try, unless he preferred adventure and an imminent death, as opposed to comfort and a longer life. The former immortals would "become" unsuccessful suicidal maniacs, if they attempted such travel, whereas the incorruptible would have to be insane to even start, and thus would not choose so to begin with.

Biblical references give us the perfect example of this kind of scenario with the angels or celestials. Angels were and are immortal, but very corruptible as is demonstrated in some percentage falling, whereas the Christ, who was Jesus, was incorruptible, but very mortal and vulnerable to physical death. Finally, we have man who unfortunately is both.

Long distance space travel, even to the nearest star system, would require both of the above, plus a little magic along the way.

With immortality, you surly would make it to any star, and at whatever speed you want. Nevertheless, maintaining your sanity forever floating in outer space, after the Doomsday particle has destroyed your craft, would be unbearable.

If there is a proper definition of a Hell, this is it. Without immortality, it would be more logical to live your simple and "perfect" incorruptible life here and now on planet earth and pass away with some piece of mind.

Since man has not obtained either of these conditions, nor gained some power of magic, interstellar space travel is, for now just a pipe dream. This must also be true for any kind of exobiological 'ETs" as well. If there were only one impossibility, it would forbid any life form getting to another star system.

DANGERS OF INTERSTELLAR PARTICLES

Interstellar ships are vulnerable to the same hazards found in interplanetary travel, such as vacuum, cosmic dust, radiation, and micrometeoroids.

Cosmic dust is a problem that needs solving before any ship takes off. Cosmic dust contributes to the radiation hazard, because the dust particles are actually lumps of high-energy nucleons at relativistic velocities. A serious problem will be the sputtering of a ship's bow or a radiation shield by the relativistic dust particles.

At extremely high speed, interstellar dust and gas can cause considerable damage to a craft. Larger objects (such as macroscopic dust grains) are far less common, but would be much more destructive. The risks of impact of such objects, and methods of mitigating these risks, have not been adequately assessed.

In every cubic kilometer of space, there are an estimated 100,000 dust particles (made up of silicates and ice) weighing only a tenth of a gram. At such a velocity, colliding with even one of these tiny objects could destroy a spaceship if it is not shielded properly.

STELLAR SANDBLAST

Let us say that the possibility of the impact of a particle over the size of 1/100th of a gram is zero. Let say, there is only cosmic dust along the way. These particles cannot penetrate the vehicle skin and are not expected to constitute a serious hazard. Yet, traveling at that speed we would be effected as if being sandblasted for 43 years.

The continual impact of many such small particles would have a gradual pitting, eroding, and sandblasting effect that could become troublesome in the case of exposed windows or optical surfaces, such as a telescope lens. This would also tax the resources of any flight with constant repairs.

As to the bigger particles, scientists say, that *"collisions with a few larger particles, weighing more than one gram, are also expected. These particles could puncture the walls of a spacecraft. However, recent data collected from experimental spacecraft indicate that the chances of a spacecraft being hit by meteors large enough to disable it are very small—much smaller than was anticipated a few years ago. Besides, various kinds of protective devices are available. These include a meteor bumper, or extra skin outside the main spacecraft skin, and several layers of thin-skinned, self-sealing materials. There is also a suitable alarm system whose purpose is to indicate both the occurrence and the position of a puncture"* [19].

Nevertheless, even though traveling at a slower speed would reduce the "intensity" of an impact, it does nothing to protect against larger impacts by particles larger than one ounce and up. It may be true that the chances of being hit by meteors large enough to cause fatal damage is very small, but it is also true that it only takes one to do it. If the meteor is big enough and there is no way to deflect or avoid it, then it means complete disaster.

As to protective devices, no meteor bumper devised is going to bump aside a doomsday meteor; no thin-skinned, self-sealing material is going to fill-in the hole, especially if the hole is half the size of the spacecraft! We may talk about micro-punctures and ways to avoid or repair them, but when traveling at such high speed, there is no protection against the big ones.   All   it takes is one miscalculated large one ounce rock. Do not forget, the increase of "speed" necessarily means the increase risk of danger and death. Plus, at 9/10ths the speed of light a 1/10th of a gram particle becomes a hydrogen bomb! Notice, 9 x 9.95 tons of TNT is now equal to 90 tons of TNT; and a one gram size particle would effectively produce an explosion equal to 9 x 2,800 tons TNT or 25,200 tons of TNT!

Speed is necessary, even if limited by the speed of light. Due to the tiny atoms scattered throughout space, any ship traveling at extreme speeds, will be impacted with such force that they would tear through even the strongest known metals.

Two options do remain. Humans or machines could constantly patch the damage. This would require impractically large amounts of repair material or the ship is made of elastic material, which self-heals. The good news is NASA has done research into such materials. The bad news is that they do not think them feasible.

RADIATION RISKS

Ultraviolet radiation: UVR is a form of short-wavelength radiation emitted by the sun. On earth, the atmosphere filters out a large percentage of the sun's ultraviolet radiation. In space, however, exposure causes the human skin to burn from 10 to 50 times as fast. Spacecraft shielding is one remedy for this.

X Rays or X radiation is encountered in space, although it is of low intensity and the walls of spacecraft cabins should provide adequate shielding from it. Alternately, secondary X rays generated by the bombardment of cosmic ray and solar flare particles on the spacecraft are of great risk.

Cosmic Rays are atomic nuclei traveling at very high speeds. They carry a positive electric charge and therefore, are deflectable to a certain extent by the earth's magnetic field.

Solar Flares of the sun and other stars also emit large numbers of particles consisting mostly of high-speed, high-energy protons, which can become very intense.

The Van Allen Radiation belt is a large, donut-shaped radiation belt. The high-speed protons of the Van Allen belt have effects similar to those of cosmic rays, and prolonged exposure must be avoided. Short-term exposure to Van Allen radiation is not likely to cause any serious effects. [20]

RADIATION AND LONG DURATIONS

For long duration missions, space radiation presents significant health risks including cancer mortality. Of these different kinds of radiation, some pose only slight problems for space flight. Others, however, constitute major hazards for spacecraft and their occupants. To categorize them, we have solar particle events (SPEs), medium energy protons galactic cosmic ray (GCR), high charge and energy (HZE) nuclei.

As to the protection against radiation on extended interstellar travels, no really "practical" or cost efficient method of protecting astronauts against exposure to cosmic ray and solar-flare particle bombardment is available. No magnetic or electrostatic shielding [alone] is, apparently, applicable for protection against nucleonic radiation of the oncoming flow, because of the presence of the neutral component in interstellar gas.

The most dangerous is interstellar gas, which acts as a flow of nucleonic radiation bombarding a relativistic starship. Radiation flux is extremely high even for moderate relativistic velocities, and therefore, proper windward shielding is a necessity. The presence of a neutral component in interstellar gas excludes the implementation of magnetic shielding alone.

Technically speaking, there are combinations of shielding efficient enough to guard against radiation, if one has an unlimited budget and unlimited resources. For instance, there are two potentially viable shielding systems. First, there is the material and magnetic shielding combination. A thick massive shield made of carbon (or Titanium) would afford relatively good (but partial) protection.

Combined with a magnetic deflection field, shielding from radiation is theoretically feasible: "*A radiation-absorbing windscreen installed in front of a spaceship...of material and magnetic shielding...can provide sufficient protection, however it becomes dramatically thicker with acceleration above 0.3c (3/10th light speed) and reaches several meters."* [21] Secondly, a water and/or ice shielding is another theoretical possibility. *"Shielding by water could also be an option... Placing a water tank (or an ice bulge) in front of a ship is advantageous... It eliminates the damaging embrittlement of solids under intense nucleonic radiation... a water tank of several tens of centimeters [10-15ft] in thickness would be sufficient... however cruising speeds closer to the speed of light would require tens of meters [65+ft] of water shielding... tons of additional load."* [22]

We said, "Technically speaking," which excludes speaking financially, for the construction of a sufficiently powerful magnet (for electro-magnetic shielding) within the limits of spacecraft weight requirements is beyond the capabilities of present-day technology.

Moreover, magnetic shielding by itself is insufficient and only partially shields against radiation bombardment. The addition of extremely thick material shielding in combination with magnetic shielding also adds huge amounts of weight, and thus too costly. This adds to the need for more propulsion fuel. The addition of such a huge amount of water is also at present impractical.

COSMIC RAYS AND BIOLOGICAL ORGANISMS

The Effects of Cosmic Rays and Solar Flares are similar in their effects on living creatures. Secondary particles are particles created by the impact of primary particles on a material. When a high-speed cosmic ray or solar-flare nucleus enters the atmosphere, the wall of a spacecraft, or any other material, it collides with atoms in the material creating particles called gamma rays. Such secondary radiation can cause serious damage to living cells.

The body apparently suffers no ill effects from short exposure to a small number of low-intensity cosmic ray or solar flare particles. However, as the radiation intensity and the time of exposure increase, there are various ill effects. Some of these, such as localized graying of the hair and the growth of tumors, take time to develop, becoming noticeable sometimes weeks, sometimes years, after exposure.

Very intense radiation can have immediate effects, such as blindness or brain lesions. In extreme cases, death may result. This radiation may also cause long-term genetic effects, which may be noticeable only in the children or grandchildren of the person exposed.

COSMIC RAYS ON ELECTRONICS

In addition to their effects on living organisms, intensive radiations have a harmful effect on certain organic materials, such as plastics, electrical insulators, and paints, and on certain electrical components, particularly semiconductors.

Radiation affects computer memory systems, which is a major problem: *"Every high-energy nucleon passing through an electronic component inevitably produces free electrons, i.e. deposits some electric charge in it, often resulting in parasitic signals and causing bits to flip, latch up, or burn out in computer memory. The deposition of charge can "upset" the memory circuits...Damage occurs when nucleons slow down and nearly come to rest at the end of their penetration depth. As a result, they knock silicon atoms out of their proper locations in crystal lattice, creating defects in crystal structure capable of trapping conduction electrons."* [23]

DANGERS AND PROBLEMS OF GRAVITY

The structure of our bodies actually depends on gravity. When humans do not live in normal Earth gravity, our bodies begin to suffer. After a few weeks or months, our bones become brittle and our muscles fatigue, with much more unpleasant long-term effects. These can be combated somewhat with various exercises and diets, but after years or decades in outer space, the human body becomes permanently damaged.

Even for relatively short flights, eyesight deteriorates so badly that NASA consider it a major boundary needed to be overcome before undertaking manned missions to Mars. Rather than living in weightlessness, acceleration from gravity can be induced by rotating the spaceship quickly. Unfortunately, this requires huge amounts of energy and fuel, and causes nausea in the short-term. The long term effects have not been studied, but are considered poor.

NUTRITION: FOOD, AIR AND WATER

Any humans living on a ship for long periods need life-support. They need to eat, drink, breathe, urinate, excrete, wash, and sleep. Many of these are addressed in space flights already made. However, on longer journeys, the amount of food and water needed becomes too large to take.

The most probable solution is to make the ship into a self-contained ecosystem. Plants could produce air, food, and be used to transform human wastes. Any ecosystem is slightly inefficient compared to earth, but it could still possibly sustain itself long enough to reach the destination – if radiation does not stunt growth and mutate the botanicals. The ship's equipment would gradually decay from the various gases being recycled, but clever maintenance or new materials might circumvent this.

The most efficient system would involve a single plant. Algae have been studied for their potential, with the "*spirspiralingae*" being looked at most closely. It would take care of air, wastes, and food. It is not a complete source of nutrition in itself, and becomes toxic if contaminated or when eaten in large quantities, but genetic engineering could change that in the future. Soylent Green in outer space?

ARGUMENTS AGAINST LIFE BEYOND EARTH

EXOBIOLOGICAL AND GEOCHEMICAL

Comparing the oldest earth rocks (cherts and microfossils) with outer space materials such as lunar regolith samples, meteorites, Mars soil samplings and Martian photography has revealed no evidence of biogenetic materials as of January 2013, and from the following examples, never will in any future date.

THE MOON IS DEAD

Lunar regolith and core sample studies show the Moon to be sterile and in most cases toxic to living organisms.

The Apollo missions collected rocks using a variety of tools, including hammers, rakes, scoops, tongs, and core tubes. The

59

samples were placed inside sample bags and then a special environmental sample container for return to the Earth to protect them from contamination. [24]

Rocks collected from the Moon have been measured by radiometric dating techniques. They range in age from about 3.16 billion years old (basaltic samples), up to about 4.5 billion years old (rocks derived from the highlands). [25]

Lunar "soil" is the fine material of the regolith found on the surface of the Moon. Its properties differ significantly from those of earth. The physical properties of lunar soil are primarily the result of mechanical disintegration of basaltic and anorthositic rock, caused by continuous meteoric impact and bombardment by interstellar charged atomic particles over billions of years. The process is largely one of mechanical erosion in which the particles are ground to finer and finer size over time.

This situation contrasts fundamentally to terrestrial soil formation, mediated by the presence of molecular oxygen (O_2), humidity, atmospheric wind, and a robust array of contributing biological processes. The term "soil" is not correct in reference to the Moon because on the Earth, soil is defined as having organic content, whereas the Moon has none. [26]

There are two profound differences in the chemistry of lunar and earth soils. The first is that the Moon is very dry. As a result, structures such as clay, mica, and amphiboles are totally absent from the Moon. The second difference is that lunar regolith and crust are chemically reduced, rather than being significantly oxidized, like the Earth's crust. [27]

A microbial assay of the lunar samples was performed by NASA scientists. Two surface samples were assayed for the presence of indigenous *kicrootganisms*. "*Neither terrestrial contaminants nor indigenous biota were demonstrated under a variety of growth conditions.*"

In fact, opposite to earth "soils" lunar dirt is sterile and lacks all evidence of having any biogenic traces. In fact, because of the higher content of calcium, aluminum, ilmenite, nickel, and scandium, most of the dirt is toxic and lethal to living organisms. Gerard Taylor said "*A microbial toxicity was demonstrated by extracts of Apollo 11 core material which had been in contact with complex nutrient media for four months"* (p. 1939).

Attempts to grow Pseudomonas aeruginosa (ATCC15442) in extracts of Apollo 11 core material that had been in contact with TGY nutrient for four months resulted in the inhibition reported in earlier tests. (See, Taylor et al., 1970a) *"The presence of this extract resulted in the loss of all colony forming units (cfu) within ten hours."* (Taylor., 1971. p. 1941). Studies done on other lunar samples gave no indication of microbial growth. (Taylor., 1971. p. 1944).

The studies done so far demonstrate *"that the fluids extracted from complexes of Apollo 11 lunar sub-surface material, which have incubated in nutrient media for four months, are toxic to all the bacterial species tested"* (Taylor., 1971. p. 1946).

Taylor's summary in "Microbial assay of lunar samples" reports finding the total lack of demonstrable, viable microorganisms (either contaminant or indigenous to the lunar material) in any of the four samples tested. [28]

LUNAR PORHYRINS

Porphyrins have been touted as proof of life. Porphyrins are a group of organic compounds, many naturally occurring. There are two kinds; those produced from biogenetic processes and those from abiotic or geochemical processes.

A geoporphyrin, also known as a petroporphyrin, is a porphyrin of geologic origin. They can occur in crude oil, oil shale, coal, or sedimentary rocks. Abelsonite is possibly the only geoporphyrin mineral, as it is rare for porphyrins to occur in isolation and form crystals. [See Wikipedia under "Porphyrins"]

Does the presence of porphyrins mean that there is or has been life on the Moon? Not at all. There are different kinds. Not all porphyrins are derived from biological processes. Some porphyrins are created by abiotic geochemical processes alone. As a group, porphyrins can form abiogenically, and are present on the Moon and in meteorites. Therefore, we have to be careful about their interpretation.

The 1969 discovery of lunar porphyrins probably said less about the chances for biochemistry there, than about how common their generation may be elsewhere in the universe. In 1978, Dr. Simionescu et al. were able to produce porphyrins under laboratory conditions similar to those of primeval Earth, before the genesis of life.

They summarized the results in the journal Origins of Life: *"Experiments with gas mixtures intended to simulate the primeval atmosphere of the Earth yielded many biologically important chemicals. Investigations into the synthesis of porphyrin-like compounds from methane, ammonia and water vapor were carried out by using high frequency discharges. Microanalyses of porphyrins showed that porphyrin-like pigments were formed in this way. The presence of divalent cations in the reaction system increased the yield of porphyrin-like pigments also involving the direct synthesis of their metal complexes. The ready formation of these compounds in abiotic conditions is significant, suggesting the possibility of their appearance during the early stage of chemical evolution."* [29]

Wherever you have life, you will have "porphryins", but where ever you find porphryins does not necessarily mean you will find life.

The reasons why the Moon is antagonistic to life forms, besides soil toxicity, are numerous. There is no significant amount of water; no air, practically no atmosphere; no ozone for cosmic ray protection; is beyond extreme temperatures; and the angle at which the moon is tilted, its speed of revolution around the sun, speed of its rotation around its own axis are other reasons.

No atmosphere means there are no aerobic organisms. It also means no protection from UV radiation and micrometeoroid bombardment. The Moon lacks a molten iron core to generate a magnetic field to protect its atmosphere in being stripped away by cosmic rays and solar winds. The Moon is not big enough and so does not have enough gravitational pull to hold on to an atmosphere. Without an atmosphere that is thick enough, the Moon cannot get warm enough, so there cannot be liquid water. Without water or moisture, life is practically impossible.

The average temperature on the surface is about 40-45 C lower than it is just below the surface. In the day, the temperature of the Moon averages 107 C, although it rises as high as 123 C. The night cools the surface to an average of -153 C, or -233 C in the permanently shaded south polar basin. A typical non-polar minimum temperature is -181 C (at the Apollo 15 site). [30]

This type of variation in temperature, and the lack of all other hospitable conditions make it impossible for life to exist or function over a long time on the Moon - unless it has a spacesuit.

The Moon turns out to be just as dead as early scientists thought it would be. *"No life, living or fossil, has been found on the Moon. The Moon is even poor in the key chemical elements that are essential for life: hydrogen, carbon, nitrogen, and others. (There is more carbon brought to the Moon by the solar wind than there is in the lunar rocks themselves."* [31]

MISSING MOON, MARS MICROBES

Have scientists found life from the moon? Russian biologists say they recognize fossils of microorganisms in the lunar soil. They reported new microscopic analysis of samples from the lunar surface, collected from the then Soviet Union's Luna missions in 1970 and 1972.

On September 24th, 1970, for the first time, an unmanned spacecraft delivered a lunar "soil" sample to Earth. The Soviet Union's Luna 16 spacecraft returned from the moon's Sea of Fertility with 101 grams of lunar regolith in a hermetically sealed container. [32] In February 1972, only 120 kilometers from the Luna 16 site, Luna 20 used a drill with a ten-inch, hollow-core bit to collect another regolith sample that was also hermetically sealed on the moon. [33]

The samples contained spherical particles that are *'virtually identical to fossils of known biological species'* in size, shape, distribution and even the way they are deformed during fossilization. They claim these fossils are solid evidence for ancient life on the moon and elsewhere in space. Other scientists disregard these claims and account them as geochemically produced. [34]

One of the problems with the alleged Martian meteorite life forms was that contamination from Earth life is almost inevitable. As to the supposed lunar microbes, this should not be a problem. The Luna samples were hermetically sealed when they were collected. However, the Russians may have underestimated the

63

abilities of microbes to invade sample containers, and no one knows for sure how effectively sealed they were during the return from the moon almost 30 years.

The origin of life from non-living chemicals (called 'spontaneous generation ', 'abiogenesis or 'chemical evolution') is chemically impossible for many reasons, even under the best conditions. The chance of life originating on the moon is slim to nothing. It has extreme temperatures, little atmosphere to shield radiation, little to no moisture, and the regolith is poisonous and toxic to life forms. If the microbes had turned out to be genuine evidence for life on the moon, this life could not have begun there.

Rather, it may be Earth life transported to the moon from violent meteoritic impact knocking earth material out, with a speed exceeding escape velocity. Apparently, the same way rocks from Mars get thrown out to Earth, rocks from Earth can get knocked out to Mars and the moon. Furthermore, moon rocks can also be ejected to Earth.

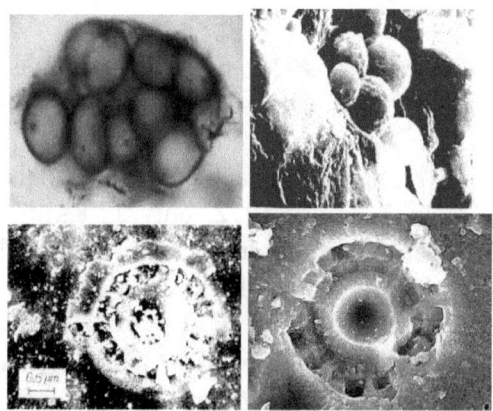

(Top Left) Luna 20: Fossils similar to modern coccoidal bacteria Siderococcus or Sulfolobus, Lithified by metallic iron. (Upper scale bar = 1.2 micrometers). (Top Right) Ancient Fossil Bacteria: Cyanobacteria from
the www.ucmp.berkeley.edu/precambrian/bittersprings.html Bitter Springs chert, Central Australia. Site date: Late Proterozoic, 850 myo.
A colonial chroococcalean form.
(Bottom LEFT) Luna 16: A silicated fossil found in lunar regolith

similar to modern spiral filamentous microorganisms such as *Phormidium frigidum*. (Bottom RIGHT) A 10-μm-diameter micro-crater on lunar glass, from Donald E. Brownlee et al. [abstract], p 1767 v 304, Science, 18 June 2004

The abiotic theory is supported by the fact that the moon fossils were 'virtually identical to fossils of known biological species' and had an 'unmistakable resemblance to modern spiral filamentous microorganisms like Phormidium frigidum, found in growing stromatolite, in Shark Bay, Australia.

Evolutionists frequently use common structures to 'prove' a common ancestry (although a common designer would explain them better), so it is difficult to believe that almost identical structures evolved independently on different places with vastly different environments.

What we may have here are naturally geochemically created artifacts, morphologically similar to biological artifacts, much the same way a micro-meteorite craters look like fossils?

SECTION – 2

IS ANYONE ELSE IS ON THE MOON!
A Critical Analysis of
George Leonard's
"Is Someone Else on the Moon"

"One has not only an ability to perceive the world,

66

*but an ability to alter one's perception of it; more simply, one can
change things by the manner in which one looks at them"*
- HENRI BERGSON, Matter and Memory

Mr George Leonard titled his book, "Somebody Else Is on The Moon." Is this statement true? Was somebody else actually on the moon? Is somebody else on the moon right now? Are there alien artifacts, construction sites, machinery and garbage dumps on the moon?

It seems that everyone researching and analyzing lunar landscapes and archaeologically digging for extra-terrestrial devices is finding all kinds of junk! In the last thirty years, this cult of alien artifact hunting has grown to pandemic proportions!

To find out the truth we must start somewhere and Mr George Leonard is the best cult leader to begin with, since his book is the first and most extravagantly, albeit verbose, extra-terrestrial travesty of selenology theories available to us. In fact, it is a master piece of macro-cosmological alien horse apple mythology ever written.

In the search for alien trailer-trash on the moon, it really does not matter where the curious ET private eye starts. Those who are artifact hunters are probably familiar with George Leonard's work "Somebody Else is on the Moon." If not, you're not a real Gung-ho alien hunter!

When it comes to lunar locations of such evidence, it also does not matter. According to George, anywhere on the moon is good enough. The moon's surface is loaded with garbage, strewn with trash and heaped with piles of junk. An astronaut alien garbage recycler would go crazy.

Mr Leonard can find ET litter anywhere on the Moon and suggests that anyone can. A re-evaluation of Mr Leonard's observations will help in the decision as to whether there really are aliens on the moon or not. Let us take his suggestion and study the Moon for ourselves. Pull up your socks and grab your Schnapps, you're going for a wild ride in search for – nothing!

HOUSTON, HE HAS A PROBLEM!

"All truth passes through three stage:
First, truth is ridiculed and then it is violently opposed. Finally,
after much time and debate,
it is accepted as being self-evident" - Anon

ET excavators, like magicians, love to take lunar photos and pull phantom artifacts out of shadows and highlighted dirt piles, crater rims and even transform boulders into lunar rabbits. The art of "deception" is to make something appear that is not there and distort what is there into something that is is not. We may conclude that all falsity passes through several stages of conflict and perversion before it is finally accepted as truth.

For example, a selenology blob may be whatever a person imagines it to be. Next, after displaying such foolishness to the public, some reasonable scientific people oppose it. Next, after science disregards it, scoffers follow mocking it. Finally, after serious violent opposition, ridicule and mockery retires back to the laboratories, the falsity (eventually) prevails from the lack of serious complaint.

From here, the lunacy spreads like a virus throughout the majority of society such as through grocery store pulp, rag-mag publications all the way to the top into government agencies. The once mocked and scoffed at foolery gains attention in the better magazines and book publishers. Money begins to flow and "talk" and the BS begins to walk its way up the cultural as well as scientific ladder - even into national space agencies.   Wisdom is, once again proven true, that the majority prefers to believe the lie, the falsehood, the deception, the myth, the legend, and the fabulous rather than the truth. Hence, the lie gains credibility from a lack of discrediting, while the truth is lost. Thus, a great woe comes upon society in saying what is false is true, and that which is true is false. Furthermore, false hopes take root and usurp true hopes.

Alien artifact hunters have the certain ability to perceive the world in an uncanny and deceptive way. By altering their rational

68

perceptions of things, more simply, they can filter vague natural lunar objects through the degenerate powers of their warped imagination and alter the manner in which they look at things, and with enough persuasion, make other lame folk do the same – see artificial alien creations. Without a solid grasp of lunar processes and considerations for poor quality images, the dim-witted reader swallows the lie of alien artifacts rather than see the reasonable "shadow" that should be creating doubt.

Mr Leonard said intelligent people want to hear what is on the cutting edge of current truth about the moon, not swallow a bunch of pseudo-scientific NASA status qua garbage (See p.16). However, according to Mr Leonard, the cutting edge of scientific truth in lunar research consists of collecting and swallowing exobiological and astrophysical pseudo-scientific "garbage" left by supposed Extraterrestrials.

According to George, the burden of proof rests upon anyone trying to prove an alien presence on the moon. Apparently, Leonard self-contracted the burden, whereas his colleagues preferred not to get their hands lunar-soiled. Therefore, in defence of ET existence and moon monuments, he wrote a book and a fancy one to boot!

In his book of 238 pages, Mr Leonard graces his readers in burdening himself to belabouring the subject of alien artifacts supposedly found as proof for an alien presence. He graces his readers as well, in that he burdened himself not to crank out 2380 pages of Selenic sewage. Further blessings abound that on his Moon there are mountainous heaps of alien leftovers and he cherry picked the best for the book.

Questioning and having healthy skepticism, as well as testing proofs are virtues in the field of science, said Mr Leonard. For, he admits all sorts of rascals might invade and lower the standards. The standards he speaks of are the standards of scientific proof. Otherwise, fictions and falsehoods creep into science and if this happen, how can we be sure that breakthroughs were real? (p. 17).

Therefore, we should be on the lookout for possible rascals infecting the realm of astrophysical sciences with trash, imprudent nonsense and alien presence artifact refuse. Yes, let us credible scientists explore and belabor these same artifacts and see what kind of burden is developed - in refraining from impertinent laughter - as Georges' breakthroughs are re-examined.

No matter his set of virtues, his conclusions are beyond incredible. He has a complete disregard for the very virtues he has set as precedence. Consequently, if the reader is to avoid the same consequence, the same virtues must be considered when analyzing the data, but without preconceptions. In this way, by using these standards of scientific evaluation, rather than disregarding them, the reader may waffle through Leonard's laughable lectures with immunity. As light exposes darkness, true science will expose Leonard's play upon shadows and highlights revealing their true contents.

Mr Leonard claims that the photos taken by the lunar missions scream with evidence that the moon has life on it. He begins bellowing loudly that, *"The moon is occupied by an intelligent race or races (and) is firmly in the possession of these occupants. Evidence of their presence is everywhere: on the surface, on the near side and on the hidden side, in the craters, and in the highlands"* (p.23).

Good Lord and lunar land sakes alive, alien rubbish is all over the moon! Therefore, it will not be difficult to blindly pin the tail on the lunar donkey and hit a gold mine of moon-dweller artifacts. Yet, with proper scientific insight, such alien rubbish will be rather only found in his book.

Without further ado, let us cast aside, as George said, all *"misconceptions and preconceptions shoveled into us"* (p. 26) and by George, and by Jove keep our minds open. We should also brace ourselves as we slam on our booster rocket "brakes" of reasoning, when we happen-stance upon alien engineering of a macroscopic scale, dwarfing anything seen on earth. Furthermore, let us be prepared to have shoveled down our throats fantastic new explanations for old mysteries on the moon, such as the artificial origin of some craters, the causeway explanation of the white crater rays (i.e. Tycho) and seriously doubt the old orthodox explanations (p.27). Let us volitionally suffer the hilarity of fantastic selenology fabrications with the full understanding that it is all alien and pure rubbish that we are looking at.

70

Is there some reason why Mr Leonard wants us to cast aside our old "orthodox" concepts and scientific explanations – basically, all sanity and decent research? There are more reasons we must do this. The first and most important reason is we can become boon-doggled, easily swallow fabulous alien interpretations, and therefore rest assured in our new religious heresy.

Another good reason, if we are privy to the myths of Mr Leonard, we can learn to create our own ET farce and parley (publish) the alien corn upon others, thus spreading the boon-doggle – and make money to boot.

Consequently, though most may waste time and money looking for lunar phantasms, fragments of asteroid Munchhausen or the Eye of Harmony, we may *"neither flux nor wither"* in profiting from our pulp fictions. Apparently, this is the case with George Leonard, for in the above noir tableau we find the photo facts do not fit the tall tail and the book is more a Daleks' search for Finn McCool's giant causeway than a science lesson.

A TRIP TO THE MOON

Let us take off in an imaginary rocket of scientific re-evaluation and fly to randomly picked spots, the first being the Sea of Tranquility. Here we will begin our reconnaissance mission into the imaginative world of George Leonard's alien artifacts, lunar life forms, and moon monuments. This is the best place to start with lunar sciences as this is where the first astronauts landed, played with feathers, hammered out lunar rocks, experimented with gadgets, and swung golf clubs.

For Mr Leonard, there must be some hidden reason other than scientific to pick this as a first choice. Of course, it is simple: Alien artifacts are there! Therefore, fasten your seat belt and stuff a

71

feather in your hat, because our spaceship is entering the lunar atmosphere and it is settling into a stable orbit around the Moon.

1
THE SEA OF TRANQILLITY

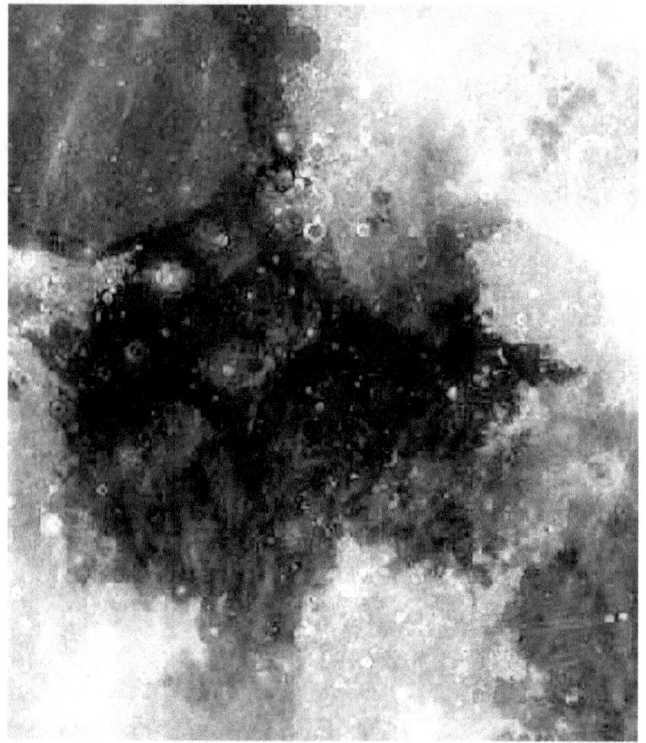

Mare Tranquillitatis
Sea of Tranquility: Named in 1651 by the astronomers Francesco Grimaldi and Giovanni Battista Riccioli in their lunar map 'Almagestum novum.'

As we direct our ship to the appointed location, we begin to see what looks like the floor of a very large ancient domed city. Now, what we are looking for are alien artifacts mentioned by Mr

Leonard and other extreme astrobiologists, if not the aliens themselves.

This is also the crash site of the Ranger-8 photographic satellite (Feb. 1965), as well as the landing area of the Apollo-11 Mission (July 20, 1969). It is also a very hefty site of alien spacecraft and constructions.

To the adventurous and scientist, the view from our spaceship is a once in a lifetime grandiose experience that cannot be duplicated on earth. It is the most breath-taking sight ever to behold short of entering heaven. Tranquility Base, here we come - again!

The panorama we see is a large area, a lunar "mare" (large "sea") that sits within the Tranquillitatis basin on the Moon. What we notice is mare material within the basin that consists of basalt and some surrounding mountains and of course, if we look hard enough, alien artifacts, domed areas, crystal lattices and roadways.

The basin has irregular margins and lacks a defined multiple-ringed structure. The irregular topography in and near the basin results from the intersection of the Tranquillitatis, Nectaris, Crisium, Fecunditatis, Serenitatis basins and the two thorough going rings of the Procellarum basin.

Concept drawing of Lunar Dome

Here we notice Palus Somni, on the northeastern rim of the mare, is filled with the basalt that spilled over from Tranquillitatis. This Mare has a slight bluish tint relative to the rest of the moon and stands out quite well when color is processed and extracted from multiple photographs. The color is likely due to higher metal content in the basaltic soil or rocks and not from aliens spray-painting graffiti or from camouflaging their lunar dwellings.

Well, shucks! Unfortunately, what we see is a disappointment, when it comes to bustling alien galactic mutant merchants scurrying about buying and selling their wares. Sorry, there are no domed cities, buildings, tollbooths, trolleybus, artificial structures, glass castles, roads, billboards or sewer pipes. All looks destroyed and what we have entered is not some Beulah land, but a hellish desolate world beaten to a pulp after billions of years of meteoric bombardment.

What are visible are the results of meteoric, geological, and morphological processes. The aliens must have relocated the artifacts to another location. Therefore, let us keep looking, for according to George, we will stumble over something. The "garbage" is everywhere and there is no place you cannot find alien evidence.

ALIEN HABITATION ON THE MOON?

Our mission to the moon went well and notes were taken as to George Leonard's claims, and the following preliminary science report made. [See "Lunar Photography and Selenographic Data Re-evaluation Preliminary Report" NASA SPAM-RSM-101]

So, where did the aliens build their landing platforms, domed houses, airports, and storage facilities? Did they move them off the moon? Did they go underground? We did not find one simple crescent wrench! Not even a threaded bolt was uncovered. Furthermore, as for one single "nut," there was no need to look any further beyond George Leonard.

NASA SP-206
"Lunar Orbiter Photographic Atlas of the Moon"

2
OVAL BLOWN BOULDERS

Lunar Orbiter II-042-H1
"The Rock Pile"
Rukl's "Moon Viewed by Lunar Orbiter (p.49)

Fret not. Trust in George. We might find some ET's. For, George directs us to a special location in the Lunar Orbiter-II photo 042-H1, which he calls "the Rock Pile".   He   claims   that UFOs are stored in a small crater among these rocks. Get your magnifying glass ready, for it is a small little area in the top right corner of the frame. In our reconnaissance of the area we found what we needed, reprocessed the photo from the original negative, enlarged it and analyzed the data, and then listed the following conclusions.

76

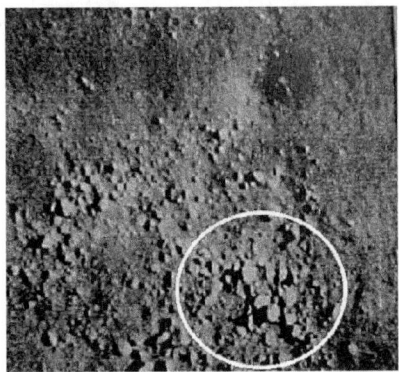

Kopal's Atlas "Megalithic Rings"

For those without access to the NSSDC archives, flip open the Lunar Orbiter Photographic Atlas of the Moon to LO-II-042-H1, and compare it to our new processed picture. You will see what George calls *"a manufactured vehicle."* He sketches what he sees on page 101, because the picture does not reveal what he said is there, even if you use a microscope. To take his word on it, he said, *"It is gleaming among other little round alien objects."* He claims the little round ovals look artificial and thus manufactured by aliens. One of the machines has *"evenly spaced rear struts"* and a *"perfect peak."* It has a beautiful *"molded point"* adorning the front, and has *"cilia like appendages, resembling those of a centipede... absolutely impossible for this kind of irregularity"* (p. 23. See Plate-2, photo 66-H-1612). Well, maybe it is a centipede and an alien one too.

Wow! This should convince and excite anyone who is bewitched into seeing in this "bad" reproduction of the rock pile an oval shaped alien artifice. However, according to George's criteria, it does not matter if you see this or not. There are plenty of other locations, in fact millions of places to look. Every NASA lunar photo is loaded with exoplasmic belching anti-gravity propulsion devices and mindless meandering digging machines.

However, at this time in 1977, George cannot know that the aliens are actually mining Helium-3, and not just pitching regolith in and out of craters. The gadgets are not digging for minerals or drilling for water, but are full of gas. George also does't know that he too is full of hot air in declaring, that this is *"one more piece of* evidence" in support of a long list "*of enormous machinery and devices that pushed the Moon around and knock down the rims of craters*" (p.23-24). It is a long list of hot air, if you ask any serious scientist.

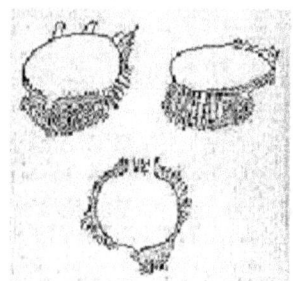

LO-II-042-H1
George's sketch of 66-H-1612. "Oval" shaped vehicles (p. 101)

Yet, George continues to point out that, "*The three 'struts' at one end of the object, with the peak at the opposite end, are so perfect as to demand the conclusion that this is artificial.*" He continues to fantasize that, "*it is a vehicle of some kind*" (p.101). In addition, he should demand us to see it, since we cannot determine what it is other than a boulder.

It is true that there are such fuzzy oval shapes. Nevertheless, shadow and light can play weird tricks on the eyes, and if the imagination is predisposed to looking for alien machines then the eyes will see alien machines. No matter what the skeptic said, the results can be convening if you are boondoggled by the alien fantasy.

Our re-evaluation of the area shows something quite different from oval shaped alien 'machines', for we see in these enlarged "detailed" photos that they are only piles of rocks, many of which form circular ring-like patterns on the lunar surface. The rings are probably the result of ancient craters undergoing deterioration and erosion. These particular boulders are actually "the Megalithic rings" and are composed of boulders, approximately 1-10 m in size. They are located in the plains of Mare Tranquillitatis (34.42 degrees E, 2.81 degrees N) as photographed by Lunar Orbiter 2, Nov. 7, 1966.

Rukl's Atlas shows an enlarged photo-enhanced shot of the area processed off the original negative (1b). Kopal's atlas also has a good enlargement of the site (1c). Most of these boulders are large and many appear to have rounded outlines. Of course, boulders can be "oval" on the Moon and can be round looking, though maybe not as smooth as earth boulders. Unlike planet earth that can smooth rocks through water erosion, lunar geophysical processes smooth the sharp edges of rocks by intense heat as they are ejected during meteorite impact or by micrometeorite bombardment over eons of time. Just because the Moon is a terrestrial type planetoid does not mean fish are jumping in and out of craters as George would like us to believe.

The following is a close up of the Megalithic Rings from Lunar Orbiter-II, but unfortunately, there is yet no LROC photos yet found of the area. Maybe you might be able to locate the frame?

3
TOOLS IN TRANQUILLITATUS

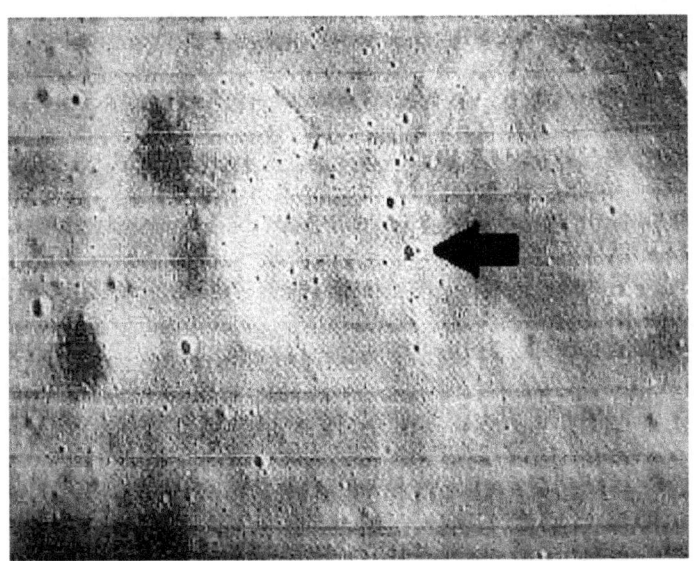

NASA 67-H-304
A sharply defined, machine-tooled object is indicated

Also in the Sea of Tranquility, south of Maskelyne-F, are other hot spots of supposed space alien activities. Mr Leonard shows another crater with *"Two magnificent machined-tooled specimens"* (See Plate 25, NASA photo 67-H-304). It is a small crater and he said, *"These creatures do love [small] craters."* Mr Leonard likes small craters too and for real good reasons. So much stuff can be hidden, stored and stuffed in "small" craters almost unnoticed.

For instance, located in the dark mare area near a broad dome of some kind is a small crater about one hundred yards in diameter (300 feet). This is one of many, he said, of the smaller more anomalous type craters to consider. Of course, what we are to see stuffed in the crater is a very *"precision made"* mechanical object with unnatural geometry that has *"a perfect square etched in blackness"* (p.177). He suggests that they are entrances to vast underground vaults, maybe caverns filled with ET provisions or even an entire metropolis.

After wasting hours scanning dozens of Lunar Orbiter frames looking for this small 'anomalous' crater, the results were just that – it was truly anomalous and impossible to find. It is a sad shame that it cannot be exposed and debunked directly, but it is not impossible to pitch it in the alien presence garbage can as more debris bunk. When compared to the billions of other empty craters it becomes highly unlikely that it houses anything else other than what every other crater contains – dirt!

Nevertheless, making an imaginative exception in this case, we are to believe it houses a *"metallic specimen"* that appears like some blanket or "lintel" - a decorative architectural element or a combined ornamented structural item like what is found over portals, doors, windows and fireplaces, but in this case a crater. It appears to be crossing over the crater, moving over the top, similar to fabric drawn like a "covering" sealing it up – a crater blanket.

Is it a canopy of some sort covering an entrance or it is an optical illusion? Since it cannot be re-locate, we cannot be sure what the lunar devil it is. George's photo is degraded and blurry, and no conclusive decision is possible. We can take his word for it or write it off as another optical illusion derived from blurry selenography.

Craters can have other weird things hidden in them besides machine tools and left over busted equipment. Accordingly, they can house alien spaceships and habitat of all kinds. These particular craters are extra small, hard to find, yet big enough to hide a spacecraft. They are also so small, yet big enough to hide fantasies! For instance, there can be diggers, dozers, dirt-spewing buggies, and most anything else we might imagine that aliens use to process the lunar surface. Craters are great places for protecting and hiding valuable items, and even excellent spots to stuff imaginary ones too.

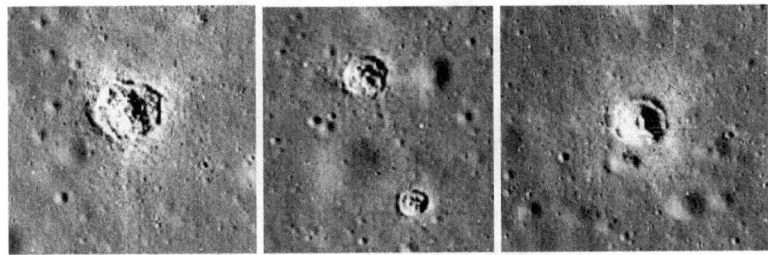

LO-III-194-H2   More craters with devices and debris.

Possible "Play-boy Bunny Rabbit" face

4
FRA MAURO HIDEAWAY-LANDS

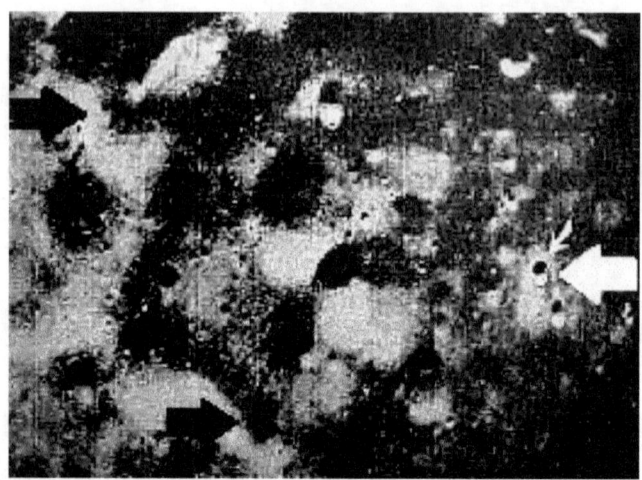

PLATE 26, 70-H-1629
Diggers, Dozers and Dirt-flingers

The Fra Mauro Highlands is a lunar formation on the near side of the moon that served as the landing site for the Apollo 14 mission in 1971. It is a widespread hilly geological area covering large portions of the lunar surface and is composed of ejecta from the impact, which formed Imbrium. It is another fantastic place to look for hidden astro-trash.

The area is primarily composed of relatively low ridges and hills, between which exist undulating valleys. Much of what covers the ejecta blanket of the Imbrium impact is debris from younger impacts and material churned up by possible moon quakes. Fra Mauro is a great hideaway for alien ejecta materials and moon-quack created craters of the same suspicion as the previous ones. Can we find any hidden anomalous alien artifacts!

Picking a crater at random and with very little discrimination is no gamble for Mr Leonard, when rooting for alien truffles. Alternately, neither is it much of a gamble for us to randomly pick one with little discrimination when exposing lunar artifices.

In Lunar Orbiter frame 70-H-1629, we have what George supposes to be some kind of "machined T-bar" shaped item hiding in an unnamed crater. He said the other craters also have oddities and activities in them. Yet, under closer inspection, these explanations themselves raise even more suspicions as to authenticity. Not one crater ever gives up any artifacts!

Mr Truffle Hunter's T-bar is most assuredly an angular rock, highlighted and shadowed in a way to look like an artificial device. Again, the crater is too small to relocate. Nevertheless, like every other crater that can be located, and there are thousands of them, it is as sure as the Sun rises that it will reveal no treasures.

Another one on the list for a one eyed and half sense inspection is a "tiny" (almost invisible) crater with what appears to be a "control-wheel" (frame LO-II-182-H1). It is very striking and looks like the head of a shank on a large screw.

George asks if this is some kind of plumbing. He speculates that it might be *an underground community.* He claims that NASA feels *if we're too blind to see these oddities* then we must have a grave problem. He obviously shares the same feeling. However, we do not share those feelings. It is not that we are so blind that we cannot see. It is that we cannot see anything even with 20/20 vision. They are not there and it is plumb foolishness to see otherwise.

Now, sane people will all agree they are probably geological oddities, but not everyone is sane. Some few will believe they are oddities manufactured by aliens, despite all evidence favoring otherwise.

LO-II-182-H1 / 67-H-041 (Plate 27)

Plate 24, 70-H-1630 Control Wheels

A close inspection reveals exactly the geological explanation as well as Leonard's imaginative ability to mold mare-mud into mare hills and paint ethereal lunar landscapes. The evidence of alien technology around the Moon is all "painted" into the imagination of his readers. Once painted or we should say "tainted," the human imagination will begin to see ET Trash everywhere it looks.

As we walk further through the black & white pages of lunar atlases in search for Leonard-e-facts, we find uncountable craters storing uncountable other extra-terrestrial crafted constructions.

As we strain our eyes, we are supposed to see different kinds of mechanical gadgets and devices. The crater in the above photo is supposedly housing some kind of "control wheel" (See Leonard, Plate 24, frame 70-H-1630).

It is a cute little crater, just like all the millions of other "cute" little craters: Natural, geologically created, full of lunar dust, riddled with rocks, and saturated with the burning heat of the Sun. However, with proper enlargement and enhancement, we see more rocks, boulders and collapsed rims with highlights and shadows.

Do you actually see any "super-rigs," "plumbing," "control wheels," "dirt-diggers," "dozers," "do-what's," "thing-a-ma-jigs," "ding floppies," and "whatcha-call-its" in them? He can fool some people all the time, all the people sometimes, but not all people all the time; and this time he is not fooling anyone anymore.

In NASA frame 66-H-1611 (above), which covers the area of a small crater west of Mare Tranquillitatis, we are to see two bumps or round objects and a "T-bar plumbing" (Leonard pl.32). Almost any small crater you pick (at random) will reveal these cute little alien bumps, pipes and bars, because almost every little crater on the moon has crater debris, rocks and boulders. So, what are they?

Did aliens wastefully leave billions of tons of their hard work behind or did natural processes leave billions of practically worthless rocks contorted into all sorts of morphological shapes? It is unimaginable that an alien culture hundreds of light years from home would be that wasteful and foolish as to leave behind precious re-usable scrap metal, junk, and hard to get resources.

Now, speaking of pipes and bars, if we get wasted enough we will see anything – not only plumbing but, kitchen sinks as well.

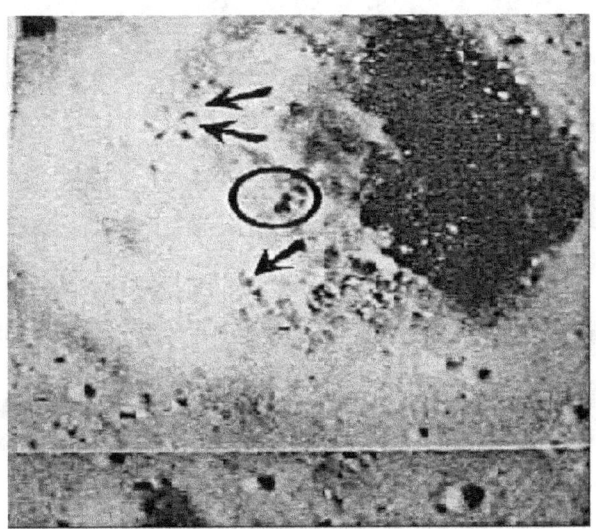

66-H-1611

5
TACKLING TYCHO TECHNOLOGY

Crater Tycho LO-V-125-M

Crater Tycho is another great hiding place for alien machines and construction sites. In fact, Crater Tycho is the ultimate hiding place, a real Disneyland of crater floor disturbances and extra-terrestrial tidbits of technology.

Tycho is located south and a little east from the great Mare area we see with our eyes on a Moonlight night. It is about 85 kilometers across and is clearly visible on our Moon's surface. The freshness of the crater and the rays of material radiating from it suggest that it is a young crater; there has been little time to erode it. Tycho appears close to the southern polar region of the Moon and surrounded by a bright ejecta blanket of rays extending across the lunar surface.

Leonard said he has found all kinds of mechanical relics lying around the crater floor *"in-between the boulders, inside the crater, outside and along the rim"* – he means everywhere. These gleaming white alien made structural hemispheres, domes, and used equipment are not real. They are irregular mounds with small craters at the peak and round mound shapes strewn about from location to location.

He said, these objects are 1/8 to 3/4 of a mile in diameter, or 400 yards, and usually never differ in size. To Leonard, this similarity in size is evidence to artificiality and the junk found is proof that Tycho is a literal alien trash dump of artifacts. There is so much alien refuse that anyone can stumble over a pile of regolithic material and find something alien!

Crater Tycho Photomosaic taken by Surveyor 7.

The central horizon hills are eight miles away from the spacecraft.

Talk about peak our interest. One location supposedly has about 30 weird objects clustered together inside it. Let us zoom in on some of these locations and see if we can find anything other than rocks, mono-mineralic fragments, various kinds of glasses including agglutinate particles, volcanic and impact spherules, high titanium basalts rich in iron, olivine, pyroxene, plagioclase, and high abundances of ferro-titanic oxide ilmenite. [1a]

Tycho crater's central peak complex about 9.3 miles (15 km) wide. (LROC)

TYCHO CRATER CENTRAL PEAK

View of the summit area of Tycho crater's central peak. Boulder in the background is 400 feet (120 m) wide.

(Credit: NASA Goddard/Arizona State University)

In the above Lunar Reconnaissance Orbiter photos, we see what looks like a "T" and "Golf Ball" on top of the central peak of Crater Tycho. This is something Leonard could never see even in the finest Apollo pictures. He did not have access to the quality of the Lunar Reconnaissance Orbiter photographs. The best he had access to was Apollo. Though very high in resolution, the Lunar Orbiter and Apollo photography does not come close to the digital resolution of the LROC pictures [1].

Further rooting for Tycho technology, George takes us to another location (69-H-1206) where we are supposed to see *"perfectly round gleaming white hemispheres,"* that are flung all about the crater floor.　　　They never vary in size, never change much in shape, but are *"symmetrical and scalloped on the straight edge"* sides. They actually rather look like chunks of pie, bitten and chewed on one side by alien pie-holes who owned them.

Nevertheless, Leonard gives us his straight answers and would have us believe they are just down to earth type structures similar to "yurts." They look exactly like the round circular yurt tents built by the Mongolian nomads of Mongolia (p.61). He said, that *"There is, however, no evidence for saying with assurance more than that they are certainly artificial"* (p.61). Begging to differ, what is artificial is certainly this kind of interpretation.

This is not all that is yurting about in Tycho: *"The floor of Tycho shows so much ground disturbance that it is virtually impossible to sort out the* [alien] *tracks from the lava flow and general ground wrinkling"* (p.103-104).　　　What?　Let　us rephrase this for Leonard: The floor of Tycho is so disturbed and contorted, and so impossible to distinguish between geological ground distortions and artificial "tracks," that it is very possible to misinterpret ground wrinkling and lava flows for alien activity, and no one would know the difference – with such blurry pictures. However, with better resolution pictures we can know with certainty exactly what they are. Let us look at some higher resolution photos of Tycho and solve this eyesore for the reader.

Leonard's drawings on page 61 describe what he sees – a vast collection of artificial whatnots. Nevertheless, he is kind enough to give us the traditional scientific (orthodox) geological explanation for them; that these mounds are the results of the upward movement of magma, which has warped the overlaying rock. An analysis of the new LROC photos reveal a positive proof of exactly what he gives as the orthodox interpretation – ground disturbances and lava flows.

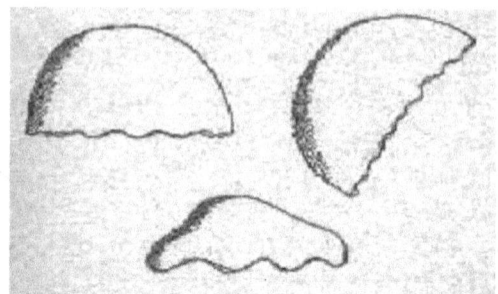

Leonard's Sketch, p.61
Some munched on pie chunks left by aliens.

On pages 61 and 124 (and plate 9) Mr Leonard publishes the Lunar Orbiter-V medium frame number 125 (NASA 69-H-1206) demonstrating what he calls domes, constructions, stitches, crater coverings and a large screw on the floor of Tycho.

The following are comparisons between his Lunar Orbiter enlargements and the new LROC higher resolution pictures of the exact same locations. You will notice immediately how easy it is to take high contrast lower resolution pictures and make ludicrousness out of

lava flows. Believing domes are homes and crevasses are constructions is just too screwy for thumbs down people too privy to such down right dumb theories.

In (A) one can see the Lunar Orbiter picture appears to present objects, but with closer inspection the LROC reveals there are no constructions nor is there any large screws in parameter (B). The supposed screw is actually a series of rough mounds that when extremely highlighted and shadowed (in the Lunar Orbiter frames) looks like a series of screw threads. The domes he speaks of that are located at one end of the lava flow rille (large crack) are actually crater ejecta. George's theory is so dome, that it should rille crack anyone up after close inspection. Here are some rille cracks and crevasses to consider carefully.

Lava flows and ground disturbances. A. Constructions and "stitches." B. Large screws in square boxes.

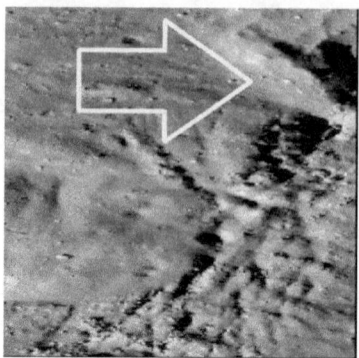

Leonard's' LO V-125-M and H2, and the LROC
Frame M129369888LC_py

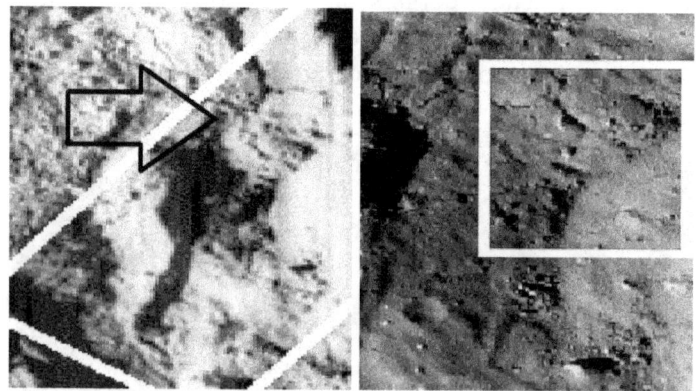

LO V-125-H2 compared with LROC
Supposed "screw" in white box is not there

6
WHITE RAYS OF TYCHO

CRATER TYCHO

The "white rays" of Tycho are a puzzle to George Leonard, maybe because he never bothered to read a lunar geology book. He said they are of alien manufacture, because they do not fit the geological model of crater development, such as a big fat asteroid smashing into the Moon. He gives a long diatribe of 10 counter arguments of why all the rays are artificial.

He said they are too big, too long, some not radially ejected from the center, many overlap and cross each other, others are tangential to it, and some end with a small crater. Nevertheless, he does admit that the rays have the same albedo as the white rocks and dirt from inside the Moon's crust. Of course, Mr Leonard, what else are they be made of?

George spends several pages trying to prove from NASA controversies over their makeup, that they are not naturally formed, and that "*the old explanations cannot be correct*" (p.108). Therefore, he gives us a new explanation for the crater rays.

The Moon of course, according to George, has been occupied for thousands of years by aliens. To explain the criss-crossing of the rays tangential to the center, he said, alien spacecraft caused them by leaving dust trails. As the UFO's fly back and forth, in all directions from the crater – for God knows what reasons, the dust sticks to the craft until it is loosed upon the surface and left behind. The dust then settles down to form all these weird directional white rays (p.110). The UFOs must have statically carried a heck of a lot of white regolith powder one minute and then somehow dumped it all as they traveled.

He goes on to quote a NASA preliminary scientific report from the Apollo-12 mission that talks about the static nature of the dry lunar dust sticking to objects, but eventually detaching from such places as the bottom of the Apollo landers.

It seems that no logical geological process can account for these anomalies of the rays. The directional nature defies all present theories - so far. They are, he said, still "*seriously debated today*" and "*we are still very much in the dark as to how the Moon's craters were formed*" (p.113). Furthermore, of all the lunar anomalies, the moon rays are the most baffling.

Google searches and photographic studies reveal that the cause of the rays are natural lunar and astrophysical processes and by nothing alien.

Dr Ray Hawke contends, "*The nature and origin of lunar rays have long been the subjects of major controversies.* [Nevertheless, in recent times, since the 1970's] *We have determined the origin of selected lunar ray segments utilizing Earth-based spectral and radar data (from Clementine, etc.)* [And found that the] *lunar rays are bright because of compositional contrast with the surrounding terrain, the presence of immature material, or some combination of the two. This fresh debris was produced by one or more of the following: [1] the emplacement of immature primary ejecta, [2] the deposition of immature local material from secondary craters, [3] the action of debris surges downrange of secondary clusters, and [4] the presence of immature interior walls of secondary impact craters"* [2]. The above definition explains the alien crater rays.

What else is going on in Crater Tycho asks George Leonard. He said NASA is supposedly covering up classified details. What we really have, according to our artifact hunter is a crater filled to the rim and brim with astro-industrial waste. George contests that alien gadgets lie on the ground built by very intelligent brains, ones at least as smart as ours.

Shockingly, we are to believe, with the less brains than a donkey, that there are things that look like letters of the alphabet and huge artificial crater coverings. We are to believe there are space vehicles, constructions of all sorts, an octagonal covering with a glyph on it, and on the inside of the rim at about two o'clock, see a long pole like object sticking out from under the lunar surface   material (p. 118-120).

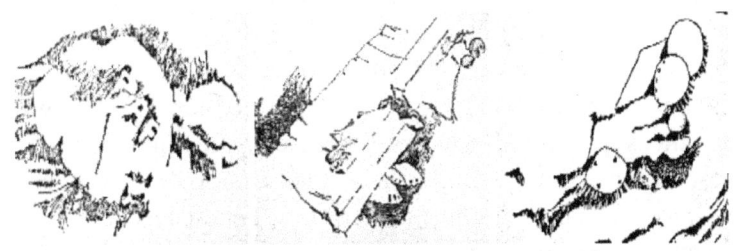

Octagonal "oval" shaped covering with
"PAF" glyph lettering (p.120)

We are to see other glyphs that look like an A, an X, and a P, and others, which are everywhere. One such object has the letters P.A.F. inscribed on it (Drawing on page 120).

If one looks harder, they may find little wooden alphabet blocks, too. The aliens supposedly used these letters to mark areas for communication, as they are found all over the moon and do suggest letter tagged location markers (p.162). This is pure bunk! They are all photographic development dust particles, hairs, and micro scratches.

Down the rim from the above "glyph" inscribed rock, we see another design with *"overall straightness,"* with the right edge having some interesting objects suspended from its rim; some of the objects have scalloping on the edges.

To the right is a geometrical cylinder-shaped object (p.122). Further detailed reveals no such objects. Here are some structures, he said, that look like oval shaped solar energy panels. The huge oval hemisphere-shaped coverings could be flat panels soaking up energy as electricity converters for the aliens.

The ovals have *"cilia like protuberances spaced at even distances on the underside,"* are perfect segments of circles or ovals, with evenly spaced 'nodes' on the rims. Some half-orange shaped blobs with cross-weavings or stitches around the edges lay nearby. George said these are remarkable objects, for they appear in more than one area on the Moon. One is a familiar item that tilts at one side, has two nodes, a cord and a bell-shaped object.

Unfortunately for George, it is also remarkable that this can be whatever the imagination wants it to be. Otherwise, unremarkably, we find some misshapen rocks, lava flows and highly contrasted light and shadow demonstrating no alien devices.

The floor of Crater Tycho is a contortion of ground disturbances as can be seen from these close-up shots. For further domes, groans, ditches and stitches let us turn to George's big screw-job, a giant digging device left behind by the aliens – obviously too big to remove.

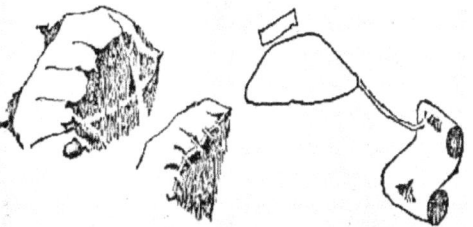

Cross-weaved stitched objects. (p. 123) and an object in the
Highlands, north of Tycho. (p. 124)

NASA 67-H-1651

More floor of Tycho Crater Tycho is littered with "alien" looking
geometric shapes, all formed out of the geothermal disfiguring of
the crater floor.

7
THE BIG SCREW

(Leonard p. 124)

There are also abandoned giant alien digging screws. Leonard wants us to believe the aliens had a huge screw job they were dealing with, and that giant twisted stitching screws are stitching together the ripped crust of the lunar surface.

Aliens could be trying to repair the Moon (a giant space ship?) with Cyclopean screws. Why this would be the case on the floor of a giant crater is unanswered. George does not care to show there are plenty of other groovy places to screw with the moon, such as all the rilles, crustal cracks, and splits (long, narrow depressions in the lunar surface that resemble channels) hundreds of miles long. These are much more in need of repair than this small area in Tycho.

Screwing the Moon back together is an enormous weight and labor even for an advanced race of aliens! On the other hand, this alien artifact is just more twisted volcanic geology distorted from twisted observations and logic.

This disturbed area is really formed by lava forced through spiral openings and solidified. Even Mr Leonard agrees that lava can be "*forced through a spiral opening*" to create that kind of shape. However, the location he said casts doubt on that explanation. The location though would be perfect for screwy looking lave formations, since the impact that caused it surly would upset subsurface lava chambers.

The following are the only Lunar Orbiter shots of a screw looking device found so far. Leonard never shows the image area, so these are probably what he was discussing in his book. Their locations seem to fit the descriptions he gives. It is the same geological anomaly from different frames. There are no LROC fames of this spot taken up to this date.

Big Screw Device -1

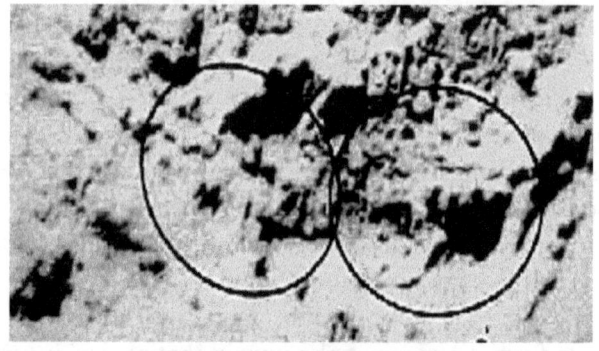

More of Big Screws

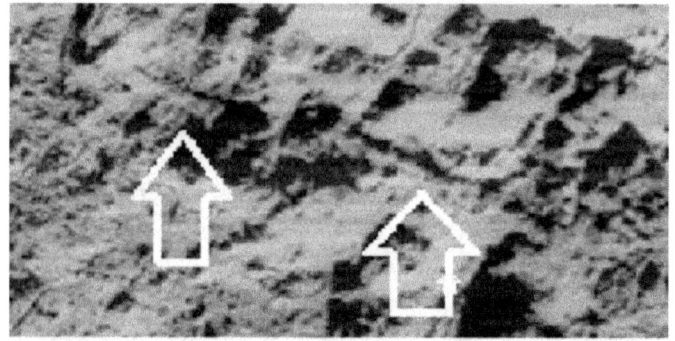

LO-V-126-M

The only big "screw" observable in Tycho

Lava can create all kinds of wacky looking contortions such as screws, twists, blobs and geometric patterns, and rectangular shapes, even square columns and blocks. Here is a fantastic lava formation that could represent alien buildings and columns on the Moon if seen from a far.

Fingals Cave, Staffa, Scotland illustrates organ-pipe type basalt columns topped by a mass of rock, the result of different rates of cooling of basalt lava when it erupted about 60 million years ago. Now, increase the size 100 times and place them in a crater and take a blurry picture orbiting the Moon, and you might see alien structures.

In the highland areas north of Tycho, Leonard said there is an area *"seven square miles"* in size..."*ALIVE with construction and activity.*" There are *"symbolic glyphs,"* *"artificial objects,"* *"some power source plates,"* some with knobs and a cord, and some

100

spaying gas, etc.　　　My God, it is a Disneyland of Alien activity! He said, it is a construction site with *"advanced technology"* that stands out vividly and *"can be taken for nothing else"* (p. 127). What else can we understand it to be? Can it be piles of odd shaped dirt and rocks?

It is interesting that the earth also has examples of such "screw" shaped lava tubes in nature. Sleeping Pele on Big Island, Hawaii is just one example of "twisted" lava flows.

(Left) Fingals_Cave_Staffa_Scotland. Organ-pipe type basalt lava. (Right) Sleeping Pele is an example of "twisted lava" flow on Big Island, Hawaii.

8
CROSSES OF KEPLER

LO-III-162-M (NASA 67-H-201)
An oblique view of crater Kepler: Location 8 degrees N, 38 degrees
W. See LROC photo:
NASA gov/images/content/513108main_012511b.jpg]

Crater Kepler is also another interesting location for galactic
garbage gatherers. It is located northwest and next to the Sea of
Insularum. It is west of Crater Copernicus and Reinhold. See
NASA Apollo-12 photo, AS12-52-7747. The blocky rim, radial
ridges, and field

THE CRATER KEPLER (LROC)
[nasa.gov/images/content/513108main_012511b.jpg]

of satellite craters are typical of young impact craters. Kepler and its related features are superposed partly on mare and partly on rugged terra that may be part of the outer rim of the Imbrium Basin.

Insularum consists of many bright RAYS or ray systems. This is perhaps the best location of Leonard's lunar domes, situated north of Crater Hortensius. Other domes lay west about the Crater Milichius that are seen with a telescope. Leonard said that these "spraying" craters are always associated with crosses. *"The crosses are perfect, (and) intersect in the exact middle, and most of the time are tipped with one end on the ground and the opposite end rose up so that they cast a shadow"* (p.65).

The angle is very oblique and raises questions as to whether the shape is actually a CROSS or is a physical anomaly caused by surface shape and angle distortion, a regular shadow and light play in the lunar mission photos. These examples of crosses, he said are not Latin or Celtic, but intersect exactly in the middle, and are usually tipped on one end with one end sticking up into the air at an angle, suggesting an alien and artificial design and placement. No natural lunar chunk of ejecta dirt would place itself in such a

103

position, at least according to Leonard. He also said there are other crosses of different shapes on the moon: "*IT (the Moon) ABOUNDS WITH THEM.*"

There is a cross near Crater Kepler, four miles in length that sticks up at an angle of 1/2 mile. It is in a triangle or square area and looks like a Roman shaped cross. The following are some of his drawings of lunar "crosses."

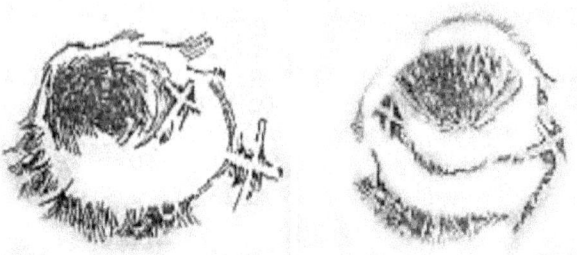

Other "crosses: scattered across the area

He said, "*Outside almost every crater being sprayed out, a gleaming cross can be seen abutting the rim. Possibly every spray crater has a cross*" (Leonard. p. 132, photo 67-H-201). Notice what Leonard said is a cross-shape sitting next the Crater Kepler is actually another natural geological formation. Nevertheless, his drawing helps to mold the mind to see what is supposedly in the photo. Yet, what is actually visible in the enlargement is an irregular shaped rock "boulder" with a semi-flat surface lying at an angle.

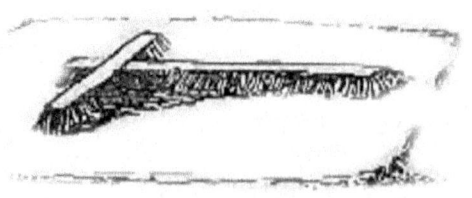

LO3-162-M
Leonard's Latin shaped cross lying at angle. (p.62)

What exactly are these cross shapes? Should we claim they are artificial religious structures before we investigate to see whether there is some natural explanation? What do NASA's lunar geologists say they are?

If we enlarge the 67-H-201 photo, what we really see is an illusion that resembles a cross, derived from a low angle sun highlight that casts a funny shadow, yet not that odd that a reasonable eye cannot detect true geology.

Look closely at the shape of the DIRT MOUND or odd shaped blob. It appears that it is just ejecta mass most assuredly ejected out by one of the surrounding impact craters. After all, the area is known for its crater rays and ejecta debris. Therefore, it is most probable that this is some sort of ejected mass rather than some artificially designed monument. The Gazetteer's reproduction of the Kepler area actually shows the extent of ejecta material and from an aerial view.

Crater Kepler has spewed out ejecta in ALL directions (like Crater Tycho) and any form or odd design can take an alien shape when it comes dropping down. Some masses fall upon others causing odd clusters, which when seen from an angle may LOOK artificial.

Other artificial structures are visible around the rim of Kepler. Alien lunar landscapers see all sorts of geometric shapes, buildings, walls, platforms, tilled fields, strip mining areas and other anomalies. Notice what we can manufacture as alien activities if we use our imagination.

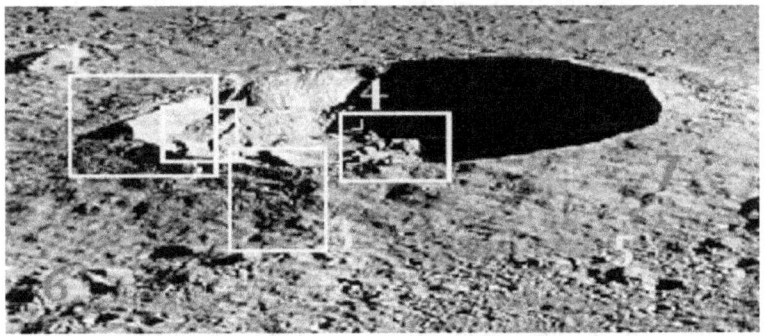

From this angle, Kepler has all kinds of
geometric looking shapes

Large flat platform

1. Shelf
2. Tilled field
3. Geometric blocks
4. Mine pits

5. More pits
6. Tilled field

9
THE BLAIR CUSPIDS
EXTRA-TERRESTRIAL TOWERS

Lunar obelisks are another oddity to arm chair geologists. What is an obelisk? An obelisk or "pointed pillar," is a tall, four-sided, narrow tapering monument , typically having a square or rectangular cross section and a pyramidal (some geometric shaped) top, set up as a monument or landmark. They can range in size and height from a few feet to 555.5 feet.

There are at least two sites with what appears as obelisks on the Moon. One site was found by the Russians and the other by Americans.

On Nov. 20, 1966, Lunar Orbiter II captured an area with six statuesque and mysterious shadows. It is a clustered area of rocks the selenographers call obelisks. They are name the "Blair Cuspids," after the researcher William Blair, who found the mysterious spires. They are in the UPLAND basin area, about the western edge of the Sea of Tranquility, near Ariadaeus Rille. The area is a wondrous place of weird highlighted and shadowed shapes and other anomalies. A meandering monument hunter can go wild in this area.

Well, now we know 'what' an obelisk is, but what does an obelisk mean? What did the ancients understand an obelisk to mean and what is so important about them?

Obelisk making evolved over a long period of time. They were first built around the time the sun cult of Re became central in Heliopolis in the 4th Dynasty.

Therefore, the ancient Egyptians were erecting obelisks around the time when they were building the pyramids at Giza. The first really tall obelisk was built by the Middle Kingdom pharaoh Senusret I or Sesostris I (1971-1926BCE, 12th Dynasty at Heliopolis. It was 65 feet tall and was cut as a single block of Aswan red granite. History shows that the ancient Egyptians perfected obelisk making over a period of more than 1,000 years.

A   **B**

(A) In the center of Saint Peter's Square, there is the 4,0year-old Egyptian obelisk, erected at the current site in 1568. The Egyptian Obelisk was brought to Rome by Emperor Caligula in 37 AD/CE. **(B)** The obelisk was originally erected in Egypt by a black African pharaoh of the 5th dynasty of Egypt between 2494–2345 BC/BCE during the Old Kingdom. The obelisk in the photo on the right is in Washington D.C.

What did an obelisk mean to the Egyptians? The ancient Egyptians didn't write a full explanation down, so we can only surmise. Stephen Quirke, in "The Cult of Ra," writes that obelisks were mainly associated with the sun. They emerged in Heliopolis at the time when the sun cult became central. Nevertheless, he feels that they might have had phallic meanings too, and that Egyptians associated both ideas, because both were considered full of creative power.

The ancient Greek historian, Diodorus, reports that Queen Semiramis erected a 130-foot obelisk in Babylon and it was associated with sun worship and represented the phallus of the sun god Baal or Nimrod. Some Masonic researchers say that the word 'obelisk' literally means 'Baal's shaft' or 'Baal's organ of reproduction'.

Pagan religions love to worship the creature rather than the Creator (Romans 1:25); and much of the onus is on the sex act, hence most temple priests were "*temple prostitutes*." Both Satanism and Freemasonry absolutely love the Obelisk. The Egyptians created the obelisk, believing that the spirit of the Sun god, Ra, dwelt in there. (3a)

Satanists and Invisible Masons depict the obelisk as the erect male organ. According to one Satanic author, "*The lingam [male phallus] was an upright pillar*" (3b) Masonry admits these pillars of the obelisk were used to represent sex. (3c)

In the Bible, we see the 'Baal's-shaft' in use in Baal worship. Listen to the account in which God destroyed the perverted pagan phallic obelisk: "*Jehu said to the runners, and to the captains, `Go in, smite the Baal worshipers, let none come out;' and they smite them by the sword, and the runners and the captains cast them out; and they go unto the city, to the house of Baal... and break down the obelisk [standing-pillar] of Baal, and break down the house of Baal, and appoint it for a drought-house unto this day*" (2 Kings 10:25-27).

Needless to say, if these lunar "phallic-shafts" are real then we have actual extra-terrestrial phallic worshiping, male-organ pillar building perverts on the Moon at least 2000 years before Christ.

The Blair Cuspids are located above Crater Cayley and a little east of Aeriadaeus B (at 15.5 degrees), and north about 1 degree above a secondary small crater (unnamed), between two tiny craters (see red square above). Using Lunar Orbiter and Clementine imagery, and especially LROC co-ordinates, they are actually located near 5.1 deg. N by 15.5 deg. E, which is within the footprint area [3].

The region of the moon where the shadows turned up is just to the western edge of the moon's Sea of Tranquility. It is an area just north of the moon's equator, slightly to the east or right, of center. This is near the area of the Apollo 11 landing site. The Cuspids were accidentally found while doing a lunar site selection survey by Lunar Orbiter II.

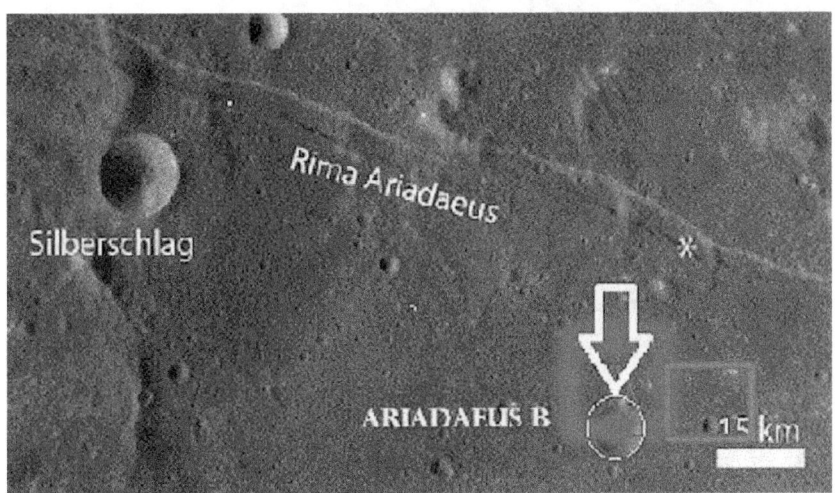

The principal photographs are LO-II-62-M, LO-II-062-H3 & LO-II-61-H3. Never mind the two "white swimming pools." The shadows cast by the Cuspids in Aeriadaeus B

Notice the group of objects with TWO bumps or raised towers. They are casting shadows like the Washington Monument on a sunny late evening! Mr Leonard noticed that the shadow of one obelisk casts over a crater obstructing the sun light on the rim.

The Cuspids consist of eight pointed spires or obelisks. They are not directly visible in the photos, but the long shadows are very visible. The largest of these according to George is 50 feet (15m) wide at the base and about 70 feet (21m) high, while the largest is about the height of a 15-story building.

NASA changed the ID numbers of all Lunar Orbiter photos. LO-II-61-H3 became the new number replacing 67-H-758. There was another image on the Orbiter's' footprint. LO-II-62-H3 included an overlap, which also showed the Cuspids! The Lunar

Orbiter-II photo ("M" or medium shot) shows the area north of Cayley and the crater Aeriadaeus B, and the Cuspids are in the bottom right half of the frame. The features stand out as bright points of light against a dark sloping lunar surface

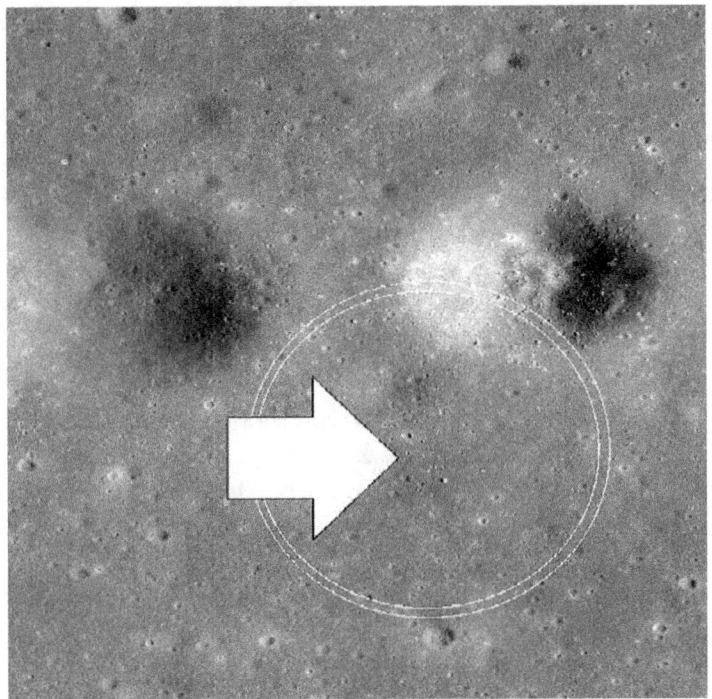

LROC M159847595RC_pyr

(The two small craters and the location of the obelisks.)
Notice in this high sun angle that there is very little "long" shadows. The obelisks are not as tall as was thought to be.

LO-II-61-H3, Leonard's 86-H-758

Enlargement of medium shot This Lunar Orbiter image shows a part of the moon surface within a crater upland basin where there are unusual large protuberances.

Mr Leonard said the objects represent an edifice of intelligent construction for living and working," or a vehicle of some kind, maybe even a form of life (p.96). Sorry Mr Leonard, they are not either, nor are they obelisks, stone phallic monuments or some form of life. They are simply large boulders as we shall see.

Nevertheless, one scientist described the needle-like shadows as the moon's "*Christmas tree effect.*" Another described it as the "*Fairy Castle.*" One scientist called the region the "Valley of Monuments" [4]. We suggest that it is just a "fairy tale."

The area presents what appears as a complicated geometry of some sort. Leonard said; "It is not enough, nor is it truthful, to simply say that the picture is 'interesting'. It is beyond the realm of possibility for the objects in it - to be of natural origin." His explanations "tower" above all else that have written upon these supposed alien monoliths.

He observes that the bigger objects "*all have two symmetrical bumps*" or "*double points*" and that the towers "*are roughly equidistant from the middle tower.*" Furthermore, the outside objects are "*all single-bumped and are spaced like sentinels.*" He said there are markings, tracks and conduits for maintaining atmosphere or communications (p.97).

112

George suggests that some of the bumps resemble "eyes", while others look like sensors. We could even take his proposition further and say that these positioned rocks look like the Cheshire Cats smile. Now, after viewing the drawing, go back up and see if you can "see" the smile without the white lines. The brain is a funny animal and it see just what it wants to see.

Are they Christmas trees? Yes, people have seen the newspaper photos and referred to them as such. George figured them as more than just trees. They were vehicles of some kind as indicated by the markings on the ground. They are devices using *"connecting cables"* and making *"tracks."*

Be careful with the imagination, for Mr Leonard warns us *"This is strictly opinion, of course - we have no right to inflict our values of design on the occupants of the Moon"* (p.98). Nevertheless, he goes on to inflict his values of design on the objects and say that *"these bumps have rounded appendages* (and) *may even be having sexual contact!"* (p. 100). It is also a good strict opinion that we should be careful not to inflict our values of design on the "geology" of the Moon – not to mention playing "gametes" and inserting sexual organelle into regolith where they do not belong.

Furthermore, other clues point to these objects as being a form of life or some mechanical substitute, because the appendages are of varying sizes, groupings and have eyes. However highly unlikely an answer this is, he said, it is more an unlikely answer that they are boulders. Well, as we shall see, it is more an unlikely answer that they are anything other than boulders!

No, they are not boulders, Christmas trees, eyes, mouths, towers or croquet balls. They are pyramids! This is what Argosy Magazine said. This rag-mag's argument goes that numbers 3, 4 and 5 are in exactly the same arrangement as the pyramids of Cheops, Chephren and Menkaura at Gizeh, Egypt. *"There are quite a number of strange material things that strongly suggest some extra-terrestrial origin or influence."* However, they are not pyramids either. Could numbers 2, 4, and 5 rather be three giant stones the size of the pyramids?

The Russians have already figured this out, George! Soviet Space Engineer Alexander Abramov came up with the startling geometrical analysis of the arrangement of the objects by calculating the angles. He asserts they constitute an "Egyptian triangle" on the moon - a precise geometric configuration known in ancient Egypt as an abaka. He said, *"The distribution of these lunar objects,"* states Abravov *"is similar to the plan of the Egyptian pyramids. The centers of the spires in this lunar abaka are arranged in precisely the same way as the apices of the three great pyramids."* [5]

Mr William Blair, who found the "stones", said that they have *"geometric patterns"* and possibly represent *"prehistoric archaeological sites like those in the Great Basin of the arid West and the Great Southwest."* He said, *"If the spires were the result of geophysical events or forces,"* then *"one would expect to find them distributed randomly"* [6]. The geometry also included a *"large rectangular shaped depression or pit,"* indicating *"four 90 degree angles"* like the profile (walls) of a pit structure.

Most astrogeologists believe the spires are the result of some geophysical event and discount the ravings of lurid lunar pyramid hunters. Whatever the 'spires" are (and we now do know what they are), they all agree that the towers are *"natural phenomena"* based on some geophysical event, and likely rock outcrops or large ejecta fragments (boulders) deposited by meteorite impacts or volcanic material ejected through faults in the moon's crust.

The structures might also be eroded cones of old volcanoes - an interpretation Mr Leonard calls *"most unlikely."* Well, finally,

114

the man is right at least once in his book! The pyramidal towers are actually not tall, say the USGS scientists, but are much shorter. The "towers" rest on a surface that is tilted away from the sun and therefore, this accounts for the long shadows. An enlargement of the photo shows the boulders are more stunted, are similar to millions of others scattered all over the lunar surface , and have no more of a geometrical mystery to them than a tossed set of dice.

The following extreme close-up enlargements demonstrate that the sun is further above the horizon, and the LROC camera is shooting at a top-downward angle and records imagery where there are no long shadows. Notice also in the following LROC M159847595RC_pyr photograph that these are only large boulders and not alien towers, skyscrapers, lunar pyramids or any other strange artificial phallic creations. They are simply MOON BOULDERS! The long shadows in the Lunar Orbiter photo images suggested these were huge towering structures, because the sun was closer to the horizon.

As we can see from the following shots the pyramids, towers, ziggurats, temples, eyes, sexual appendages or giant triangle markers are only huge boulders. See the following extreme close-ups.

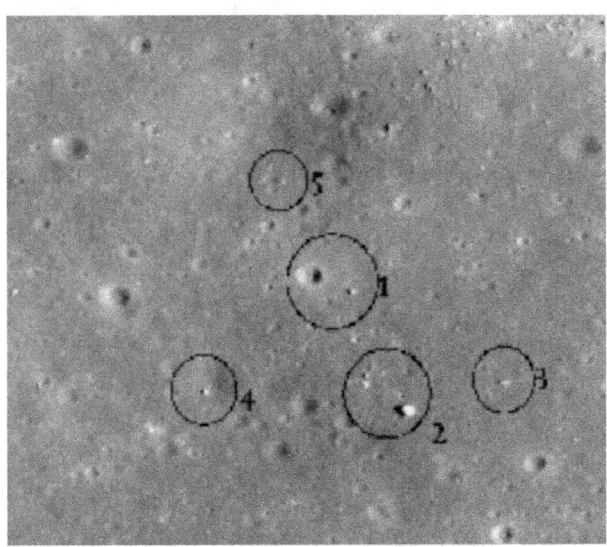

LROC M159847595RC_pyr

BLAIR CUSPIDS (Closeup-1)

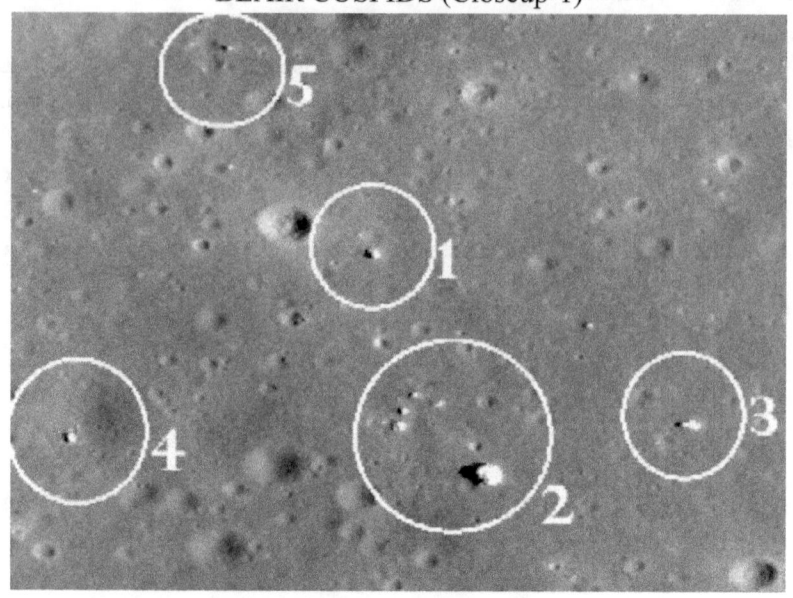

LROC M159847595RC_pyr

Extreme Closeup-2

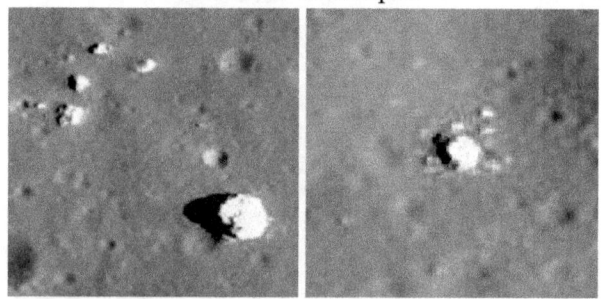

No. 2 Rock   **No. 4 Rock**

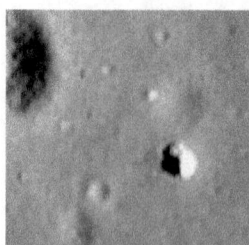

No. 1 Rock (Extreme close up)

10
THE MOSCOW MONUMENTS

Notice the sun is at a very low angle, causing the object to cast an unusually long shadow. Notice all other long shadows in the photograph! Note also, as LUNA 9 continues to take Photographs, the sun rises changes angle and the shadows decrease in length. As the sun rises further the shadow decreases more exposing the objects true size. The boulder is not as "tall" as some claim.

The Russians say they found a site full of obelisks built by aliens before we did! It seems that no matter what Americans come up with, the Russians need to have their version and they usually claim they are first.

118

The Russians acquired a set of photos showing obelisks in the Sea of Storms a few months before NASA captured the image of the Blair Cuspids. They are about two thousand miles away from the American site.

These are the Russian Cuspids. Of course, we see that this began a great storm on the subject of what these objects were.

The Soviet probe Luna-9, after its soft lunar landing in February 1966 photographed a series of three panoramas depicting an odd set of stones that appeared to be all the same in size, shape and geometric formation.

The obelisks found by the Russians also presented a strange geometry similar to the American obelisks. The "towers" cast very long shadows suggesting the objects are very tall and thus unusual for lunar geology – even though the!

The three photos presented here show this shadow cast by a small solid object, one of a number of unnatural looking "markers" staked out across the lunar surface by ET's. The photos show an unusual arrangement of stones, many of which are the same size, shape, and set at identical distances apart.

After careful study of the LUNA-9 photos, the Russians concluded that the distance between stones 1, 2, 3 and 4 is equal. The stones are identical in nature and in measurement, they said. They further claim there does not seem to be any height or elevation nearby from which the stones could have rolled and scattered into these geometric positions.

The objects appear to be arranged with definite geometric laws and mathematical patterns. Some alien intelligence must have arranged them according to definite geometric patterns for a particular purpose.

Leonard claims to prove that the cluster of bumps in the American photo, display a real geometry, just as the Russians claimed for their obelisks.

The monuments are arranged in such a manner that the relationship between the outer objects and the inside cluster are equidistant from the inside mid-point. Leonard mentions that at the time of release, the Russians believed they were artificial and even published an article relating the associated geometry as unusual for lunar ejecta.

Both the American Lunar Orbiter II photographs and the Luna-9 pictures were widely published in the Soviet Union, whereas very few people in this country even consider them. Russian scientists have always been extremely interested in the pursuit of any evidence of extra-terrestrial life. They are now asking whether intelligent beings could have visited our moon long ago and erected these permanent monuments. The LROC photos answer this question.

Are these "obelisks" of alien origin? Who built them? Maybe the Chinese Emperor Yao and his scientists built these obelisks. Maybe ancient Hindu scientists got there first and erected them. Perhaps the Druids build them as a lunar version of their Stonehenge. Maybe the Egyptian gods Ra and Amen, in their heavenly sun boats went to the moon and built some big types of wankers like the ones in Egypt. Maybe it was someone's grandmother. How about maybe one of those Arabians on a flying carpet? The other scientific possibility is they are natural lunar geophysical results of crater ejecta - "boulders" tossed out of impact sites after meteoric bombardment.

LUNA -9 IS STILL LOST

LO-IRP Restored version of LO-III-214-M
Restored image of Planitia Descendus, possible landing site of
Luna 9.

The following image was taken by Lunar Orbiter III in
February 1967. This oblique photo shows the region around the
crater Galilaei and Planitia Descensus in Oceanus Procellarum (the
Sea of Storms). In the upper center of the image you can see the
Great Wall of Procellarum. (Lunar Orbiter Image Recovery Project.
See MoonViews.com)

Lunar Orbiter Image Restoration Project (LO-IRP), captured
this shot of Planitia Descendus (the low hills and ancient craters on
the peninsula at the near center along the western Oceanus
Procellarum) restored from the original Lunar Orbiter III (LO-III-
214-M) in Feb. 1967 [7].

Somewhere in this image is the landing site of Luna 9, the
first successfully soft-landed vehicle on the Moon, sent by the
Soviet Union and arriving the same month a year earlier, on
February 3, 1966.

121

The Luna 9's landing site is disputable, because the Luna 9 images show a flat region with what may be hills some distance away, though official coordinates (296.0° E, 7.0°N) put the egg-shaped 82 kg lander right in the middle of the rough terrain of the mountains of Descendus. There is obviously not much flatland at that location. So where is Luna 9? There have been many excellent, educated speculations over the years.

Just this side of the largest distinct crater on the flat plain of the Sea of Storms, at the top of this image (west of Reiner Gamma, out of view) is the resting place of Luna 8, which may have landed successfully Dec. 7, 1965, though all contact with the earlier Soviet probe was lost at or near landing [8].

The Russian Luna 9 spacecraft came to a soft landing on the moon February 3, 1966 in the area Planitia Desscentus "Plain of Descent," a borderland along the western edge of Oceanus Procellarum. No one is sure the exact location, but there are possible candidates, while others are not. The best, but still inconclusive spot is located at 8.48 degrees N, 64.47 degrees W. in an area that covers approximately 14,300 square kilometers. The possible landing spot is one of only a few locations with one focusing on an "object of interest" at sunrise (Photos 1 and 2) when the sun was less than 5 degrees over the east horizon. Notice at this time of morning the long shadows allowing for depth perception. In the following LROC photo, we see more than one long shadowed object (Photo 3).

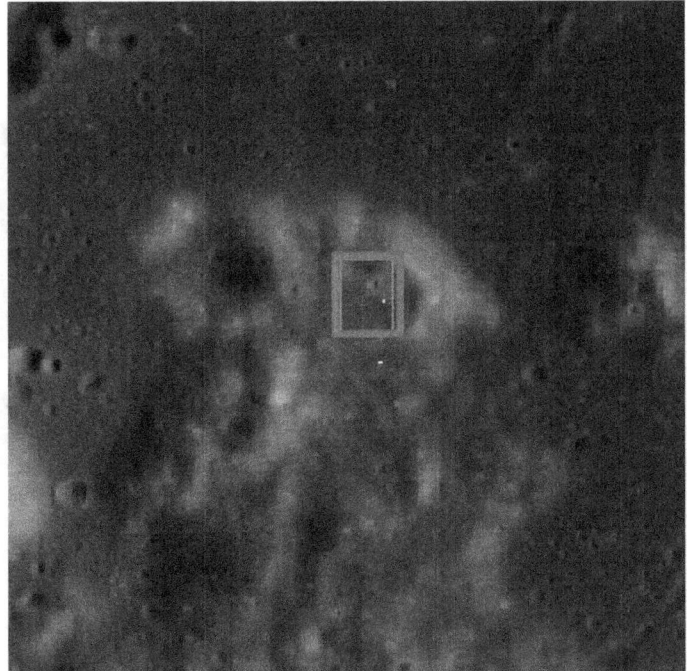

(Photo 1) Supposed Luna-9 Location M132071202LC_pyr
Planitia Desscentus "Plain of Descent," borderland along the
western edge of Oceanus Procellarum.

The following shots are taken from, enlarged and enhanced
from LOC NAC M132071202L orbit 4597, taken June 25, 2010 at
high resolution. The arrows point out possible locations. The white
circle "1" identifies the most probable "object of interest." The
object certainly casts a shadow, as anything would when caught in
morning sunshine on the moon. It is also a bit brighter than the
other similarly sized objects (2, 3, 4 and 5) caught in this 238
meter-wide field of view.

The object appears to be sitting within the west slope of a 10-
meter crater with a darkened interior. Since Luna 9 landed not long
after local sunrise, this does not help identify the lander. Another
bright object to the southwest (Object 2) is too large to be Luna 9,
but its size and location to the circled object in the crater makes a
case for it being part of the larger lander bus [9]. Object 5 is also
too big, while objects 3 and 4 might hold some future possibilities.

123

These might end up being large crater ejecta boulders making this is a wild goose chase.

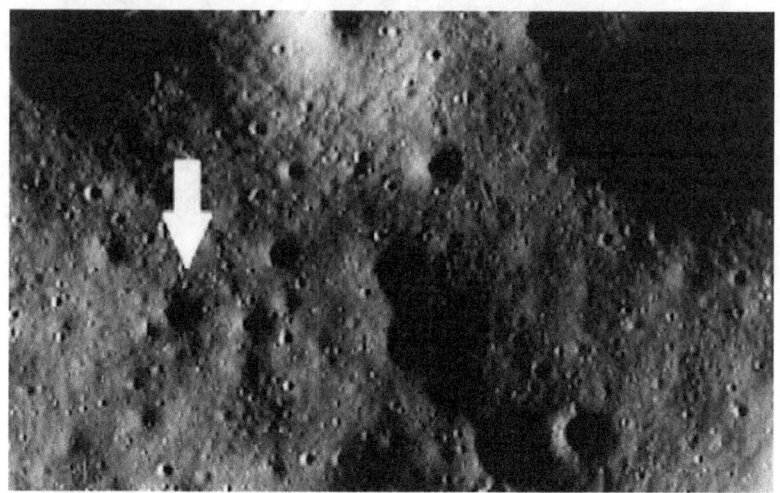

(Photo 2) M132071202LC_pyr

(Photo 3) Possible area of Luna 9 landing site

OBJECT-1 Luna 9   OBJECT 2 Lander Bus

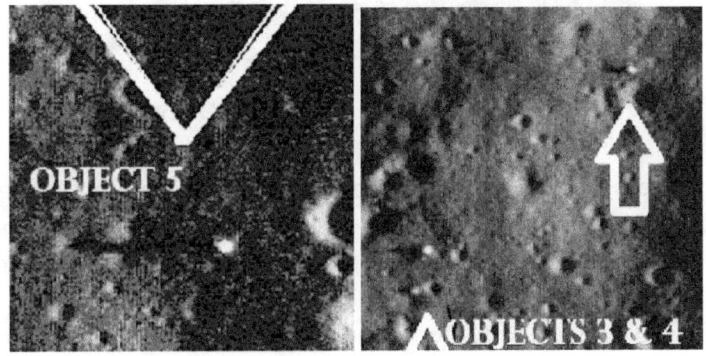

OBJECTS 3 and 4   OBJECT 5

M132071202LC_pyr Enlargement

11
THE L-BOW SHAPED OBELISK

125

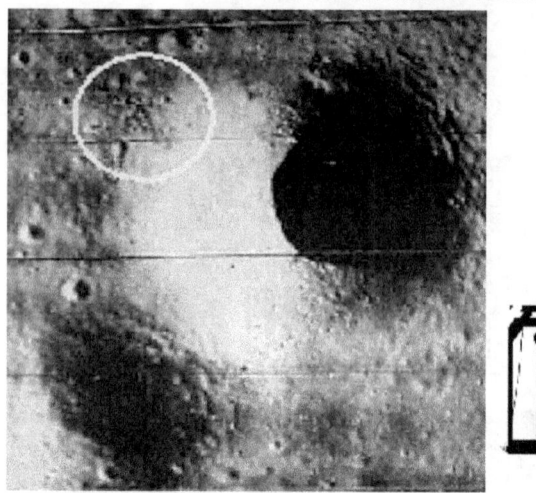

**67-H-187 (III-012-H1)   Leonard's Drawing p.169
reinterpreted.**

If it is not plumbing, mechanical legs, rolling dirt flingers, it is obelisk elbows! The following photograph was taken on Feb.15, 1967 and shows something rising about 100 feet and then for no geologically known reason, turns and juts out a horizontal BAR at a perfect 90 degree angle. Our friend George shows us his drawing of what he believes it looks like from photo blow-ups. Actually, the following photo presents an interesting set of objects that appear to have some symmetry and artificiality.

George calls this object one of the *"sentinels in the wasteland"* and a good example of a single tower *"rising straight up from the ground – not on peaks or highlands."* He continues that in some cases there are towers spaced several miles apart and perfectly aligned.   His favourite one is the above LO-III-012-H1 "L-Tower," which rises for perhaps a hundred feet or more and then turns suddenly horizontal at a perfect 90-degree angle (p.168).

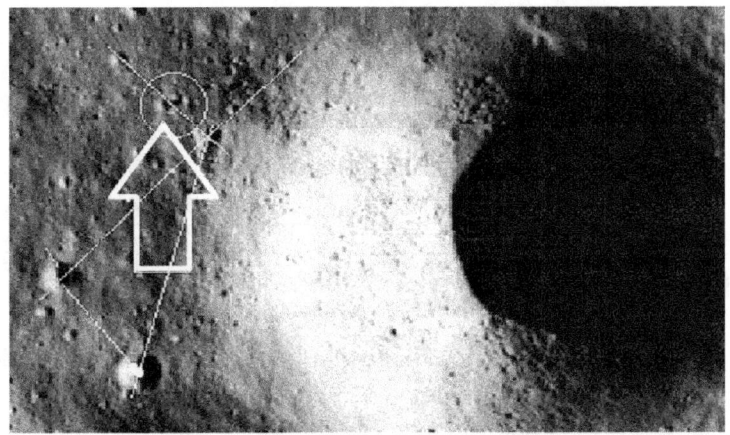

LO-III-012-H1 High resolution

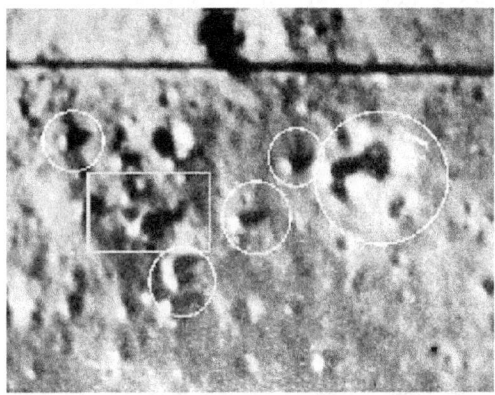

Leonard's 67-H-187
PLATE- 21 L-BAR" is in square. Other "towers" in circles appear
to be surrounding pyramidal structures to complement the "L".
(Kopal, Pl.308, p.280)

The area (Longitude 35.27 degrees, Latitude 2.71 degrees)
shows a cluster of stones, where one tower has a horizontal jut. It
shape is like the letter "L". The object casting it (in the yellow
square) looks like a mechanical star shaped 'thing', which is drawn
and mapped out. The photo has a directional arrow pointing up,
letting us know Leonard published the photo upside down.
Nevertheless, the following extrapolations derive from the lunar
Stonehenge.

First, we draw a grid with perspective lines. Secondly, we draw the lunar surface and use a 30-degree angle for sunlight. Then, chart out perspective lines and add objects and shadows according to the grid pattern, the shape of the rocks, and the structure and morphology of the lunar surface. The lunar surface in the area has crater ejecta fields and mounds of ejecta running among these boulders. One main ridge runs along the tip of the "L" shaped shadow. This causes a false interpretation of the top of the object.

Looking from another angle, we see the tip of the "L" shadow rises up onto a raised surface area. The topography of the area is not perfectly flat, but irregular. At about 30 degrees, the sun is casting a shadow from the rock that makes it look taller than it is and the tip of the shadow runs "up" and against a raised area, which helps to bend the top. Yet, the top of the rock probably does have an odd shaped "jut" to give it that "L" look. However, it is not as "geometric" and tall as one would think.

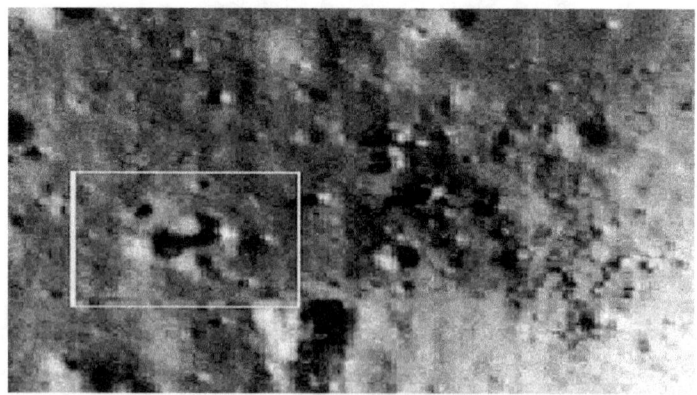

"L" Shaped Tower

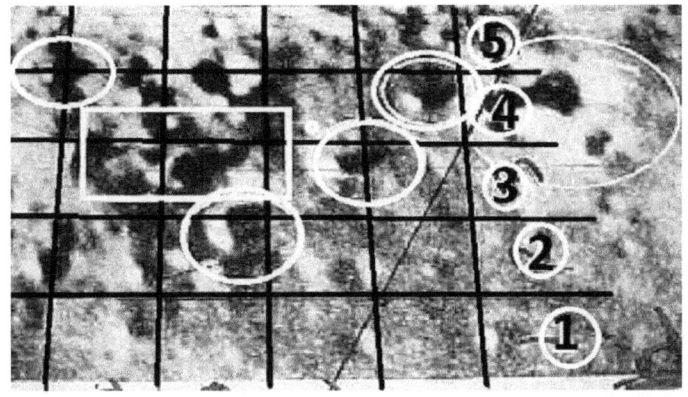

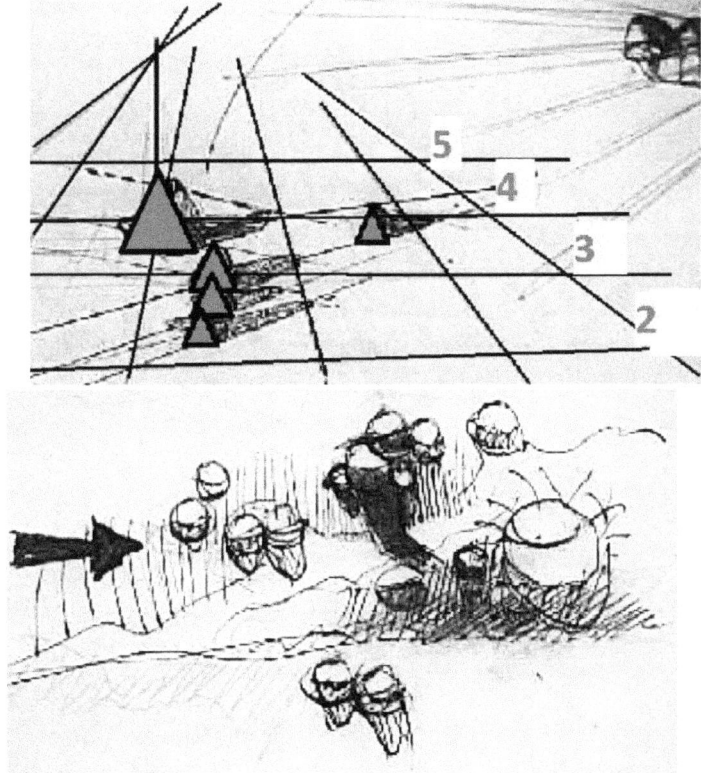

Top View

129

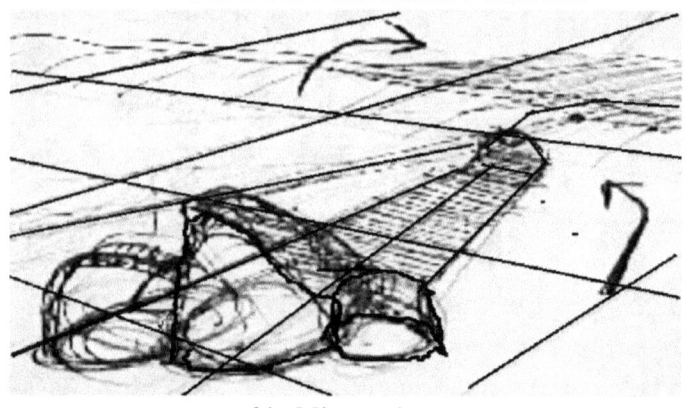

¾ oblique view

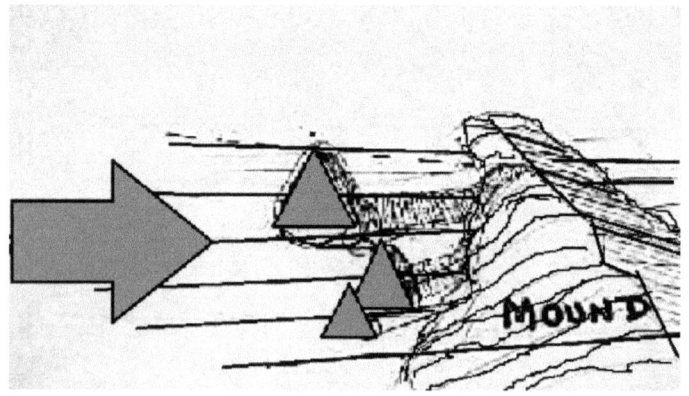

Side view

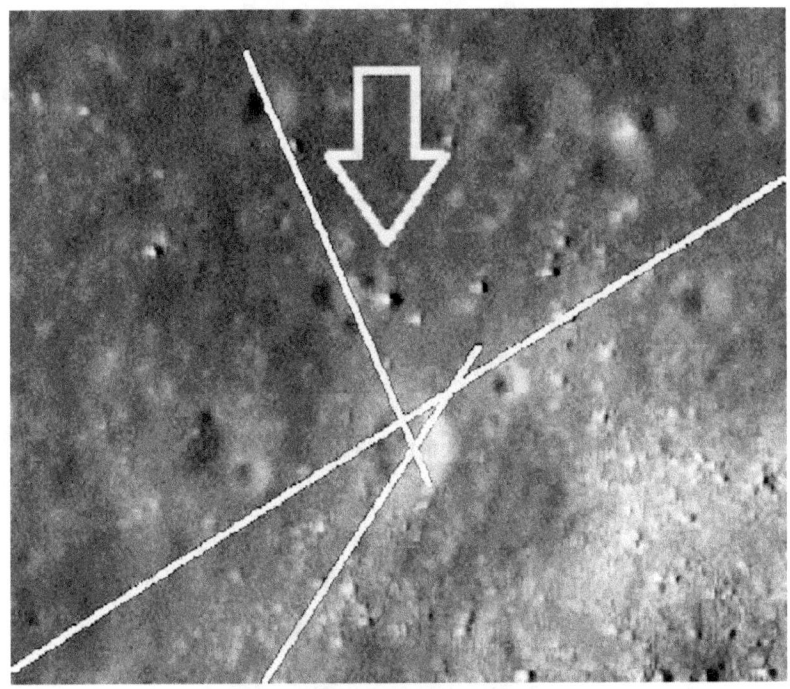

LROC M1098623356RC_pyr
"L" Shaped Tower or "boulder"? See any "L" shaped obelisk?

12
OUT OF "CONTROL" WHEELS

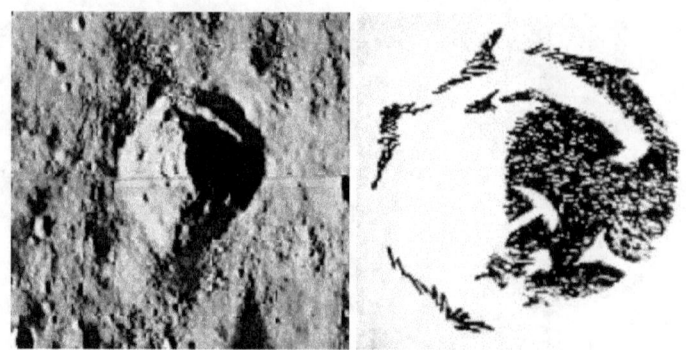

LO-II-182-H1 (NASA 67-H-041)
PLATE 27 Control Wheel in crater.
(Drawing p. 179)

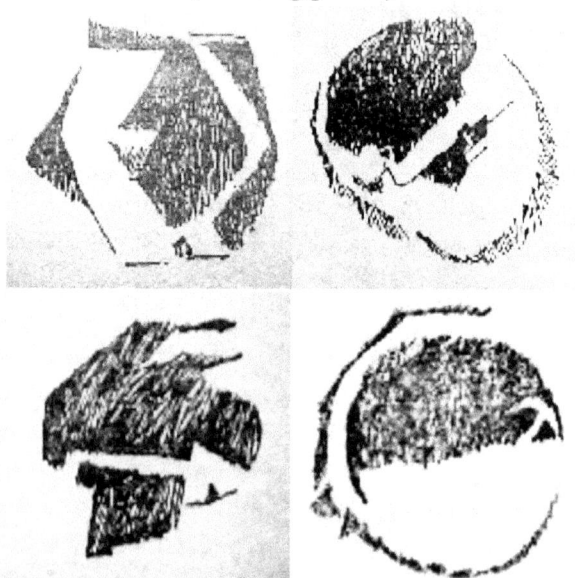

Machine tools.
T-shaped pointed device. Diamond shaped tools.

One of Leonard's' favourite habits is to throw a moon artifact at us, write up a page of imaginative speculation and then refer us to a NASA photo supposedly supporting the description. Unless you have superhuman eyesight, you have to trust Leonard's out of control interpretations. Such is the case with the following mechanical "control wheel." (p.145)

132

Southeast of Crater Kepler there is a site with lots of smaller craters and in one of these, George said, has some weird mechanical thing jammed in it. This is the Kepler Crater ejecta site in (67-H-041). The ejecta area around the crater is full of oddities like many craters on the Moon. He said there is one with a "control wheel" in it that seems to have some kind of "shank" or "screw" attached to it. He suggests it is some kind of plumbing.   We have to beg the question further, for only crater rocks, debris, and collapsed crater rims exist. Here are some examples where George jams all kinds of mechanical things into craters for our enjoyment.

After slamming NASA for teasing us with obvious alien artifact photos, George directs us to other photos teasing us with even more details.

This Apollo photo shows more "control wheels" if one can see that close. However, without enlarging the crater, Leonard draws it for us (p. 179). This "spoke wagon-wheel" is near a small crater in the Fra Mauro area. It looks like a wheel lying flat from an airplane's view.

Is it an entrance to an underground community? Is it a doorway to the hollow moon? Maybe this wheel opens some hidden door. On close inspection, it is shadow play along a collapsed crater rim and Mr Leonard is wheeling and dealing a big spoof. A twist here and a twist there, add some fantasy, and POOF! We have an extra-terrestrial steering wheel the size of New Your City.

Humbug, Mr Leonard! If your grandmother had wheels, she would be a wagon too! It looks like one of the thousands of other craters with collapsed rims. Here are some examples of wheels and other oddities in small craters. The material found in the craters is crater debris and nothing else.

13
CRATER RINGS, DOMES AND SPEAKERS

LO-III-181-H3   III-204-H1   III-181-H3

There are not only craters with spoke wheels in them and pieces of plumbing pipes, but there are craters with double rims called "double craters." These are craters with what appears to be a

134

smaller crater on the rim of a bigger one. They can consist of double impacts or collapsed crater rim. The following are some examples:

[Ref. Moon Viewed by Lunar Orbiter (p.69)]

168-H3   III-118-M[2]   III-118-H2

Gruithuisen K   125-H1   AS8-12-2052

We have seen craters with double rings, convex and concave craters, and crater domes. So, how about crater CIRCLES? In previous examples, we saw stones that look like obelisks, stones that look like Stonehenge, stones that roll around, and stones that appear arranged in circles, even stones that roll up and down hills. Some of these sites look like scattered beds of boulders cleared out in the center in certain areas, creating meadows, with a perimeter of stones jumbled around them.

Here we have another geological anomaly that represents craters with rings and round mounds inside. The ringed craters look like giant "speaker cones." Could the aliens be listening to us? They are concentric craters, ones with more than one crater rim, caused by double impacts or other volcanic geological processes.

The double ring or inner "second" ring (as depicted in the above picture of Gruithuisen K crater and the succeeding four photo frames) is more likely the result of successive volcanic eruptions, but could be the result of slumping around the entire wall [10]. A combination of both processes may also configure such boulder layouts.

135

What about domed craters in LO-III photo 118-M? Could they be some alien domed habitat structure? The origin of convex floors or crater mounds are rather thought to have been produced by elastic rebound through stress waves reflected by competent strata or discontinuities beneath (Schultz. P46). As to alien listening devices, it is becoming evident that listening to Leonard's devices is becoming more incredible.

14
BLOBS AND BLACK CROSSES

In photo 67-H-935 (LO-IV-187-H2) we have what appears to be an alien construction site (Leon. p.146 plate 29) abandoned 1000's of years ago, or at least right before the Lunar Orbiter took the picture.

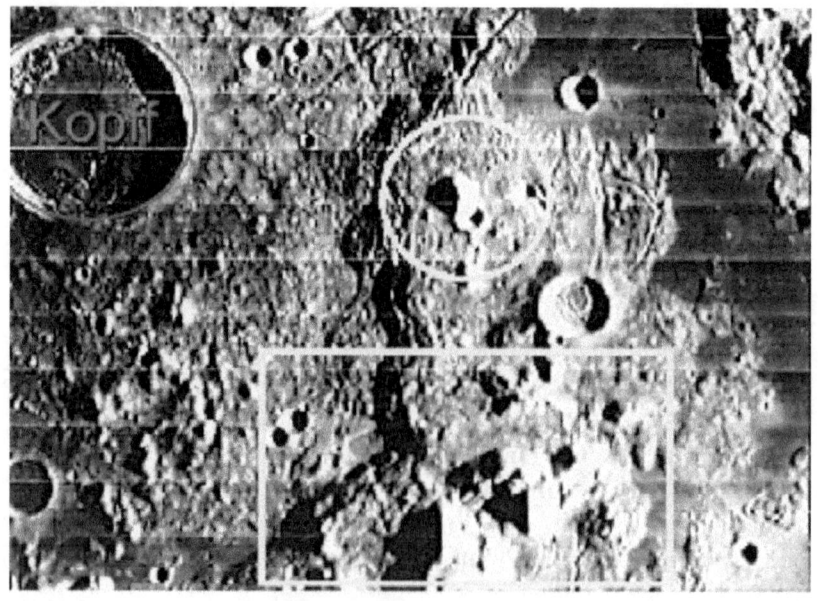

LO-IV-187-H2 (Leonard pl.29)

This site is located in the areas of Mare Orientale, Mare Veris and the Rook Mountains. The Oriental Sea is stuffed full of lunar geological oddities. It presents in extreme close-ups and enlargements every possible geological events that could take place. It is a geologist's dream world of exciting shapes and geophysical processes. There are cracks, crevasses, ridges, mounds, bumps, lumps and lava flows to mention just a few. In the upper right of Leonard's photo, we cannot help but see a rock with some kind of black cross, stamped on it.   Here at the Orientale site under construction we also find a crater with some blob "mound" covering 1/2 of it. Just strolling about Mare Orientale is exciting enough, even without the alien constructions concept. There are tons of weird (alien) looking constructions scattered everywhere.

There are more oddities here for the geologist to ponder on than Carter has liver pills! For example, LO-IV.187-2 has all sorts of "things" sitting around on crater rims and in valleys between the hills. It is a wonderment why Mr Leonard did not compose an encyclopedia.

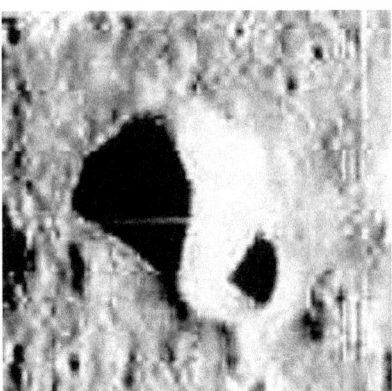

Close-up of frame LO-IV.187-2

Hummocky terrain has numerous mounds, blobs, cylinders and other odd shaped geological features. This one isolated peak (in the above photo) happens to be sitting over part of an older craterlet. They are more likely crater ejecta that have landed on a

crater rim or are volcanic in origin. See Steckling section 2
concerning "domes" for a complete analysis of this object.

LO-IV-187-H2

These are some of the close-ups of the area. Black crosses,
domes, odd shaped mounds, patterns, spheres and shelves, cracks,
crevasses and craters.

Are these decent explanations? Leonard gives us his
description of the area. See if you can see these things. He said
there are parallel walls with an arch between them with sun shining
beneath the arch.

There are nodes, raised markings at exact symmetrical spots
on one wall, and *"each node (is) on a line with the inside line of the
two walls, each an exact distance from the corner, each with the
same size shadow."* Could these alien towers actually be crater
ejecta boulders? What has the highest probability of being true:
crater ejecta boulders or alien artificial objects? After just a short
survey of lunar photography, we have found nothing. If the moon is
littered extensively with extra-terrestrial trash as Mr Leonard said,

we should find at least one conclusive example of proof. Let us continue, for galactic garbage may not be all that common.

LO-IV-186-H3   (Black Crosses)   IV-187-H2

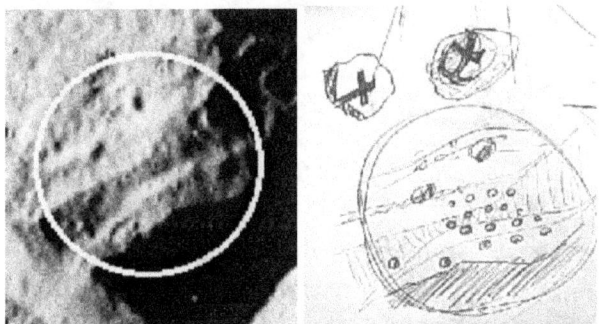

IV-186-H3   Pattern of "spheres"

15

PROCELLARUM PYRAMIDS

AS15-M-2087, 2088, 2089

How about pyramids on the moon? We have seen possible domes, roads, pipes, crafts, machinery and supposed obelisks. So why not pyramids? George Leonard's photo, p.147, plate 30 [from NASA frames 71-H-1300 and 1765, AS15-M-2087 to 2089] is a good example of what could be Egyptian like structures. The three tall and one small pyramid-like towers stand out above all other surrounding objects. They cast long shadows as compared to the other "mounds" and "blobs." These tall piles of planetary pyramids are taller than the previously discussed obelisks. They are located (at the right edge of the photo) of the Herodotus mountain range and Oceanus Procellarum.

Why are these areas interesting to Mr Leonard? He finds in them every kind of construction imaginable. The area, he said, is loaded with *"dozens of mountain masses, a long valley, many constructions and domes."* Obviously, these are alien habitat!

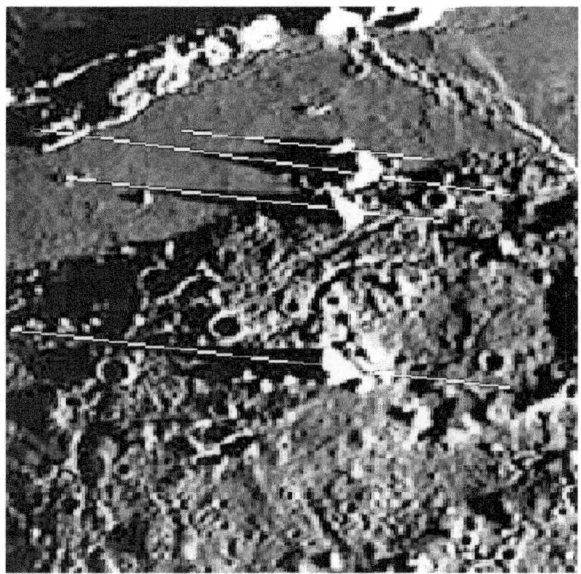

AS15-M-2087, 2088, 2089

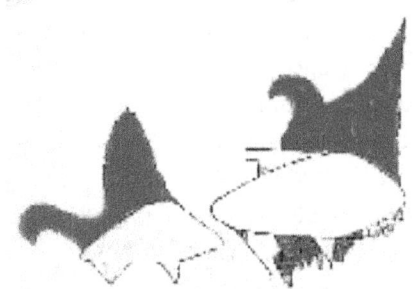

Leonard's drawing of the "domes" p.182

He said that the domes are not natural as those created by *"volcanic swelling"* or mantle stresses that are usually *"low and irregular,"* and often adorned with a small crater at the summit. These domes are true mounds and therefore artificially made.

If artificial, then why round and why are they many times built in or next to small craters? NASA has the answer, if Leonard does not. In support of lunar habitation, craters are perfect for possible living quarters.

They are round like a quarter, have a rim like a quarter and are somewhat flat like a quarter. They just as well are "two bits" as domes or pyramids. Nevertheless, they may serve as already made

141

fox-holes to bury the rooms. The builder does not have to spend extra time and alien currency at digging. Mother Moon or the 'Man in the Moon' has done it for us. It is half way already built for us – or the aliens.

This kind of shelter is also good shielding against outer space dangers such as radiation. Radiation on the Moon is very different to that on the Earth because the Moon lacks both a strong magnetic field and a thick atmosphere. Thus, natural radiation protection on the surface of the Moon is non-existent. Radiation hazards on the Moon come from two sources: Galactic cosmic rays (GCRs) and solar energetic particle events that produce extremely high levels of radiation.

These events are serious radiation hazards and the heavy GCRs and the secondary particles require the use of extensive shielding [11]. Moon craters are perfect foundations for this! All that is needed is a little regolith cover-up and George has done a real good job at this – covering up the real facts about ET foxholes.

It is not far-fetched to say that the environment of the habitat influences its design. Regolith (lunar soil) is the perfect shielding for radiation and the "dome" shape is the perfect geometric structure for such a dwelling; and NASA is in the process of designing these type domed structures for the Moon.

Leonard would have us believe that the moon-monument makers beat us to the punch. Whether they have or not, NASA is engineering dome structures as the perfect shape for lunar habitation and considering putting them in craters. This surly reminds us of those large domes sitting on crater rims and inside craters.

Extreme environments, whether they are the frigid nights of the Polar Regions, the burning heat of the desert or the harsh environment of Selenia, the goal is to provide radiation protection, while also providing an aesthetic living environment.

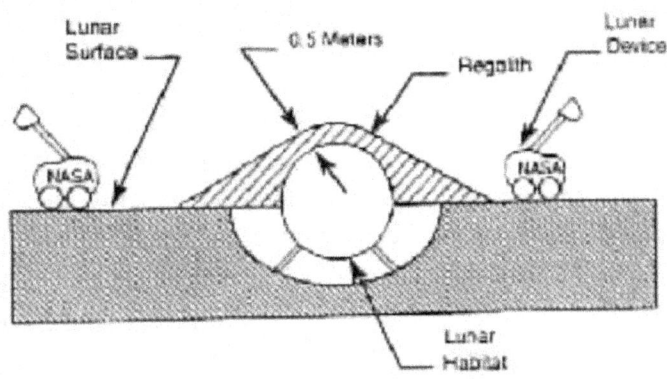

Domed habitat using crater for protection.
("Domed Cities..." p. 4)

NASA said, *"Because of the need to provide for radiation protection, a unique structural system... was created. The system uses cable networks in a tensioned structural system, which supports the lunar regolith used for shielding above the facilities. The system is modular, easily expandable, and simple to construct. Additional innovations include the use of ... various sized craters to provide side shielding. The reflective properties of the fabric... are utilized to provide diffuse illumination. The use of craters along with the suspended shielding allows the dome to be utilized in the Moon's hostile radiation environment... construction techniques for large domes...100's to 1000's of meters - for the design of habitats for long term use in extreme environments"* [12].

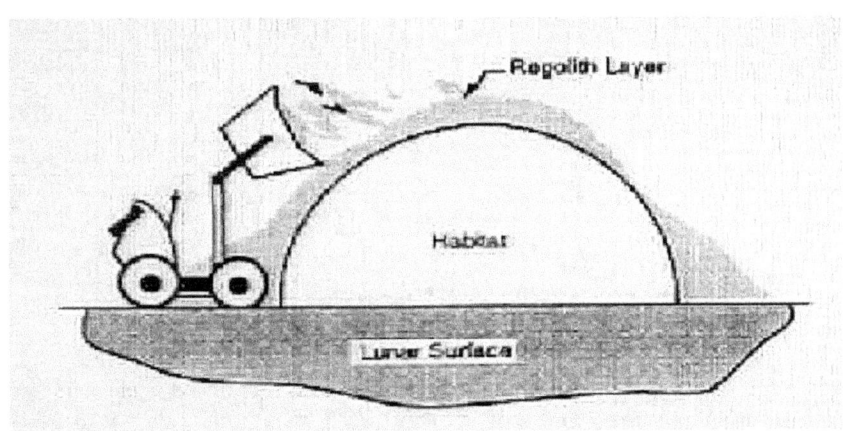

143

Regolith "lunar soil" covering dome for shielding purposes.
("Domed Cities..." p. 20)

Do we see alien domes in craters in the enlargements of the frames? Closer examination reveals nothing of the sort. There are no Genesis Moon camps, alien airports or crystal domes. When blurred and cheaply published, the lunar photographs shadows and highlights offer an interpretive frenzy to the imagination of dome hunters.

When properly examined in detail, these shadows and highlights reveal a clustered jumble of rock piles, lumps of crater ejecta and mounds of dirt with shadows and highlights playing "cloud formation" tricks on the eyes of the selenographer. Now properly exposed, Leonardeans should return to their clustered pile of NASA prints and re-examine the lumps, bumps and lunar dumps they call domed houses and reconsider their fancies. They must also answer why the aliens built many of their domes on the "rim" of the craters and not inside. Maybe they missed calculated layout and placement and missed the mark?

Speaking of getting high on the highlands and finding more of these lunar domes, let us now turn to the Alpine Mountains.

16
ALPINE CONSTRUCTION, CO.

LO-IV-115-H3 (See also LO-IV-H-116-H1)

The Alpine Valley is located on the near side of the moon and on the extreme northern edge of Crater Plato. The Alpine Valley looks out upon a vast mare or plain area, while in another direction lie mountains. This is the region of the surface dwellers according to George.

145

In moon maps and charts, you will notice a big crack close to Plato. It looks like an old scar after a decade of healing. There are supposedly constructions near the Rock Mountains and Schickard Crater in the southeast region. George has come up with the crackpot idea that the above ground builders built domes on platforms around this area and beyond the Ocean of Storms.

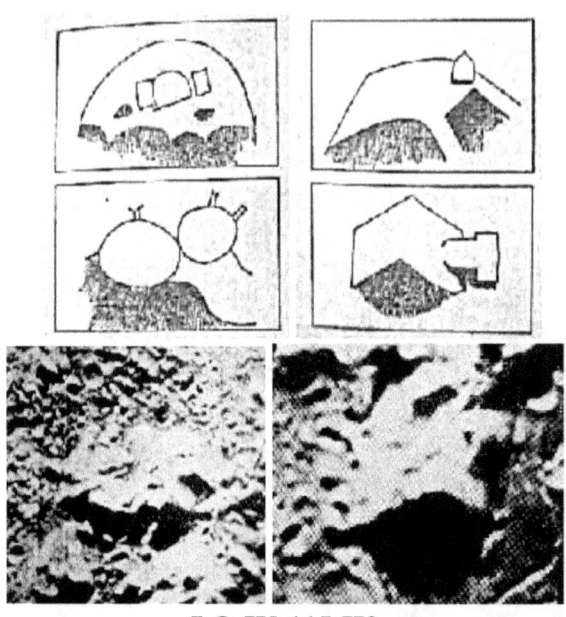

LO-IV-115-H3
Dome, top left; 'sawhorse' dome top right.

Leonard finds many interesting constructions lying around this area, as indicated in the above picture. He calls it the Disneyland of the Moon. In one area he sees an odd shaped "domed structure" (p. 187) sitting on a platform and identifies it as "an abode" rather than a spaceship (p.188). There are many "oval shaped" objects and other "circular" areas scattered about the area suggesting that many of these are dwellings, living quarters and platform style terraces.

Almost all of them said Leonard are not natural but artificial structures - "*Platform after gleaming platform...*" each with its own brand of similar size dome and approximately one half mile high,

146

and some six miles wide (p. 189). They are so big (six to ten miles wide) that you could stuff a small town inside one.

Mr Leonard could have imagined hydroponic gardens and recreational areas as well. Here are some of the structural shapes in the area identified by him as alien dome habitat, platforms and other dwellings.

The Valley formed by volcanic and erosive geological events in the early development of the Moon. Leonard said that the alien occupants keep watch on us with their telescopes from this area. To him, alien artifacts pitched from balconies clutter this whole area. He said, "*A residence of one of the habitats high up on a sculptured platform has a view out on a broad mare, or plain. In another direction there are mountains, and everywhere there are the carved aesthetics so tied to his existence...they live above ground in domes and other sculptured geometrics carved on top of gleaning flat platforms.*"

George then asks whether the following drawing of a weird artificial object "sitting on a platform" intrigues you. He calls it a dome and not a spaceship, because it is funny shaped and is not changed in the photos, taken months apart. In the area, there are ovals and circles rising to beautiful geometric peaks. He goes to conclude that NONE of the mountains and platforms is of natural origin. All the platforms range from 6-10 miles across, while all the domes are about an equal size of 2 miles.

IF you look hard enough, which Mr Leonard neglected to do, you can also find a sculptured "duck" and what looks like a huge animal "skull."

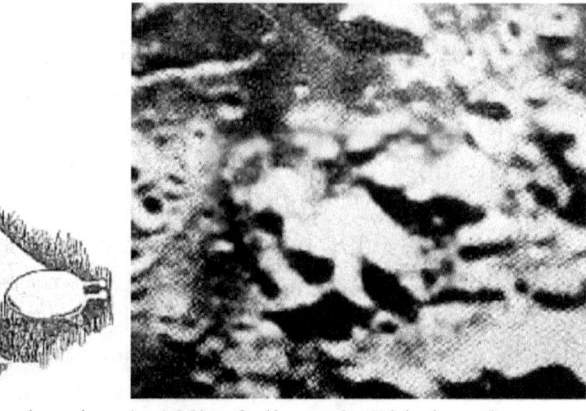

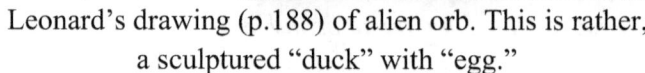

Leonard's drawing (p.188) of alien orb. This is rather,
a sculptured "duck" with "egg."

The duck sculpture seems to be intact as opposed to the other,
and apparently has some "egg-shape" nested next to its tail, which
is Leonard's odd shape dome. The skull's "right" ear has collapsed
from decay and erosion, while the nose area has fallen to its left
side. After reconstructing and attaching the dilapidated remains to
their proper locations with computer graphics software, we obtain
the formal original shape of the icon or sculpture of a mouse face.

The only connection found between aliens and rodent-duck
iconography, which Mr Leonard also neglected to research, lies in
human history. It is obvious that the ancient lunar inhabitants were
observing our ancestors, and if the 'duck" and "mouse" icons are
contemporaneous; they may still be doing so - the duck icon dates
this site as modern. It could be that later (surviving) Selenian
generations connected the modern "duck" to the more ancient
mouse icon, being that the duck carving is much less eroded and
thus appears more recent.

Why mouse and duck imagery on the moon? What could a
mouse mean to lunar inhabitants? I shall attempt to give an
explanation based upon the postulation of alien presence.
Apparently, if aliens have been on the moon they have been there
for a very long time.

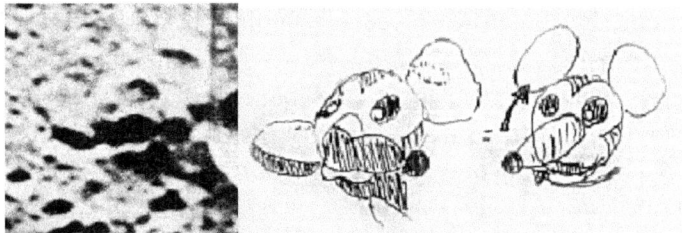

The "Skull" or renamed "Mouse Rock."

Rodent iconography found on the moon would indicate the artifacts approximate age relative to human history, placing such relics about the time of ancient Greece and possibly back to at least the last dynasties of Egypt. There were rodent iconography and ancient mouse cults in Greece thousands of years ago, as well as in Egypt, India and other places.

Alternately, there are no such formal mice cults (religions) in the modern age, yet some entertainment images contain the vestiges of zoomorphic veneration. Mr Leonard's suggestion that areas of the Moon are Disneyland-like was prophetic of these newly found discoveries.

Besides the mouse (*vamana* "vehicle") association with the god Ganesha, we have those of the Sminthean cult of Apollo. Historians are unclear as to the exact connection of the mouse with Apollo and are not sure what Apollo's original mouse-smintheus association was. Nevertheless, there are allusions relating to a possible connection to disease and the power to control it. Mice seem to control disease by being carriers.

Smintheus could be a geographic name derived from Sminthia, an ancient city in Asia Minor, which Strabo said is near Hamaxitus. Homer places other early cult centers of Apollo Smintheus in the cities of Chrysa and Tenedos. Depictions of Smithean Apollo show the god standing with a mouse under foot. He is also shown with a mouse in hand.

The only relationship between rodents and the god Apollo is that both caused and averted plagues. It seems, according to the ancient Greeks, that any god who could control mice could also control disease, pestilence and starvation, as well as defeat in battles. Apollo Smintheus protected farmers from field-mice

149

depredations, he therefore staved off famine, pestilence and thus plagues.

Here's what Strabo has to say about the connection of the sminthean mouse (Strabo XIII). *"The temple of Apollo Smintheus is in this Chrysa, and the symbol, a mouse, which shows the etymology of the epithet Smintheus, lying under the foot of the statue... They reconcile the history, and the fable about the mice, in this following manner. The Teucri, who came from Crete,... were directed by an oracle to settle wherever the earth-born inhabitants should attack them, which, it is said, occurred to them near Hamaxitus, for in the night-time great multitudes of field-mice came out and devoured all arms or utensils, which were made of leather; the colony therefore settled there... Heracleides of Pontus said, that the mice, which swarmed near the temple, were considered as sacred, and the statue is represented as standing upon a mouse"* [13].

So far, no evidence of "living" aliens is found on the moon, though tons of suggestive astro-archaeological evidence is supposedly surfacing to the contrary through photographic analysis. Nevertheless, the possibility that aliens visited earth, contracted diseases and viral infections, and even brought them back to the moon is highly suggested in this lunar iconography. Could it be that the aliens, just as Apollo had done, tried to harness the power of Apollo through colossus size mouse iconography (talismans or Amulets) to battle diseases? This icon and the fact that no living aliens have been found, suggests that the Selenians were wiped out by pan-lunar plagues.

The plot thickens: Whence the duck? Could the duck represent survival, celebration and Vanity Fair? It is tempting to think that the plagues were as water off a ducks back and that they have survived to this day, maybe hiding this time underground.

The photographed area under close inspection actually reveals that the above hypothesis is a regolithic load of alien Poo, based on the same hyper-dimensional over the hill synthetic thinking of the Leonard type mind set. With a little wine bibbing and some crackers and cheese, it is not hard to "dream" up ideas out shadows cast from moon light, like a small child spooked in his bed after the lights are turned out - especially after watching a good horror movie.

The area amounts to the same contrasting, varied and congested lunar geological formations. The platforms are just huge areas with bright spots; the domes, either ejecta or swelling uplifts, while all the other "constructions" of supposed ducks, mice and other such zoomorphic fictions are only varied and assorted lunar morphologies, radiating shadows and highlights in a Disneyland of geological deformations.

151

17
CAUSEWAYS IN CRISIS

AS16-121-19438
Apollo shot of northern rim of the Sea of Crisis

Dr Farouk El Baz, another alien artifact supporter, who taught the astronauts geology, said there are spires on the moon much higher than any constructions on Earth. He also mentions funny unexplainable and unnatural flashes of lights: *"There's no question about it, they are very tremendous things: no comets, not natural."*

152

They were also seen by Ken Mattingly on Apollo 16 and by Ron Evans and Jack Schmitt on Apollo 17. (Leon. p. 150)

Ivan Sanderson, another early alien theorist of Agrosy Magazine, stated flatly that the Moon is littered with constructions everywhere. Many Russian Scientists, at the time the photos hit the market, saw all kinds of wild things in the photos. Joseph Goodavage (a prolific writer, journalist, remarkable astrologer and humorist) believed he saw alien artifacts, artificial constructions and ancient ruins in lunar photographs! Most of the larger and more prominent artifacts are located specifically around MARE CRISIUM - the Sea of Crisis [14].

Mr John O'Neil, a science editor for the New York herald Tribune reported in 1953 that he saw (through his telescope) *"a gigantic natural bridge having the amazing span of about 12 miles from pediment to pediment"* [15].

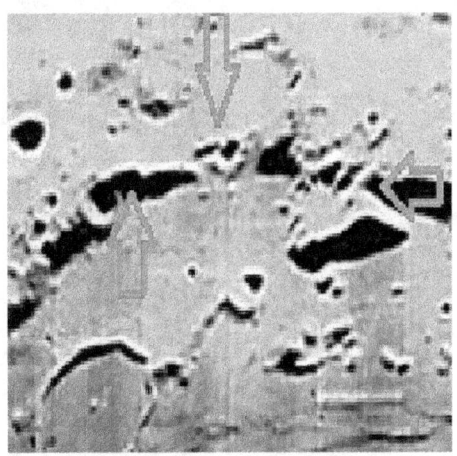

LO-IV-1191-H3

THE CRISIUM CAUSEWAY

O'Neill observed his bridge at the thin neck along the western shore of Mare Crisium just where two promontories come almost together. The rim of a large ruined crater to the left, a smaller bright crater, and two ridges in the mare probably appeared as arcs of shadow and brightness, giving rise to the interpretation of a bridge.

He said the main bridge between the two promontories was *"straight as a die"* and cast a shadow beneath. O'Neill's Bridge

153

was popularized at a time when amateurs thought anything was possible, including UFOs. Percy Wilkins, in his 1954 book, "Our Moon", confirmed the existence of the bridge and hinted that it could be artificial. This, and other embarrassments, was too much for the other members of the association and Wilkins was forced to resign as Director.

Nevertheless, bridges apparently do span huge fracture lines in what geologists refer to as scarps, such as can be seen near Crater Lalande, 50 miles west of Sinus Medii (AS16-M-0849).

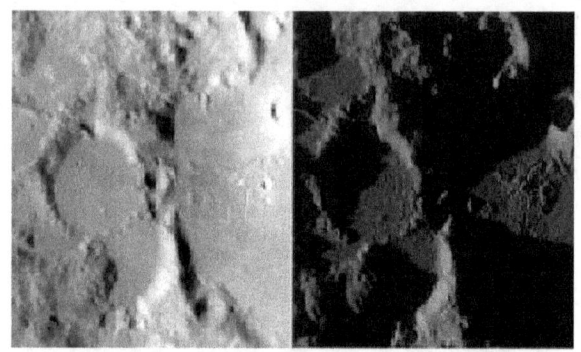

Mare Crisium Promontories at different times.
(Compliments of http://www.the-moon.wikispaces.com/)

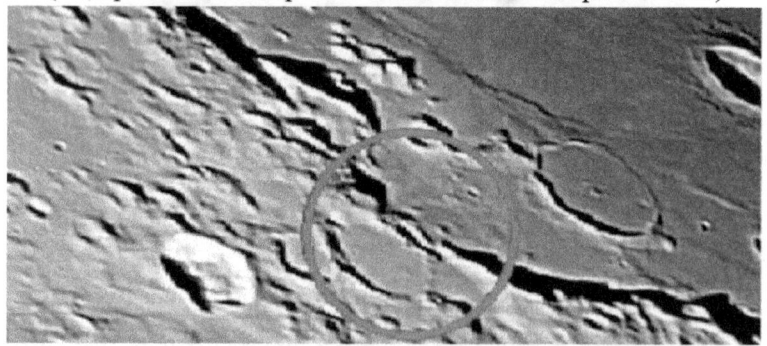

Mare Crisium "bridge"
(Compliments of Alex E. Norman, 'Anacortes Astronomy Club')

The "bridges" appear as two narrow filaments crossing the black expanse of the scarp's trough. Areas of the plains on both sides of this scarp are covered with patterns of parallel and linear ridges, some making sharp 90-degree turns. The general appearance is similar to a high-altitude aerial photograph of a city, connected by bridges across a canyon to the outlying suburbs [16].

Twenty-four years later, Mr George Leonard stumbles upon these same "bridges" in his telescopic observations, while Moon cruising for alien activity. He relocates the bridges that O'Neil saw decades earlier, running along the northern mountainous ridge area of the Sea of Crisis. The bridges he saw were huge.

Mare Crisium is located in the Western Hemisphere of the near side of the Moon. It sits above the equator and looks like a large SEA area, or plains.

It is part of what we see making the face of the "Man in the Moon." To the North and along the lower rim edges of the old crater rim of the Northern Highlands one is supposed to see what appears to be bridges spanning over from one area to another. It is a jumble of assorted geological mishaps, erosion and weird ridges.

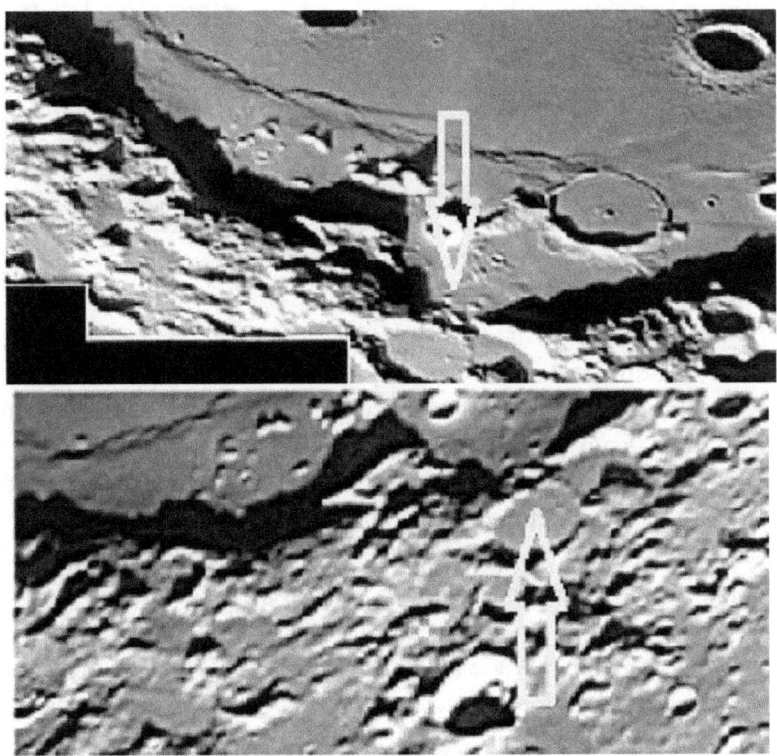

Mare Crisium "bridge."

(Compliments of Alex Norman, 'Anacortes Astronomy Club')

It truly is "the Sea of Crisis" as far as the Highlands are concerned. George said, *"The entire area is filled with constructions of various shapes which rise into the sky. Some are bent over, not touching the ground. Others touch the ground and become 'bridges,'* (plate 1)." To clarify his observations, he presents what he thinks he sees. Here are some interpretations and the close-ups to match.

The Sea of Crisis

Nevertheless, it is also a major crisis for the reader to see what he claims. After showing my local astronomy friend what I was looking for he managed to capture some close-ups of the bridge. Yet, from the above NASA Apollo photos just about any ridge, raised area or highlighted odd shaped crater rim can be made into an alien manufactured artifact or bridge.

If one looks long enough through a telescope, he will eventually see what appear to be O'Neil's bridges. It is hard to catch and capture, but it can as the above pictures prove. Nevertheless, what we see are not bridges, skyscrapers, high-rises or decaying crystal domes, but sun light playing off crater rims. We see a beautiful view of the point where Prom Lavinium and Prom Olivium almost touch.

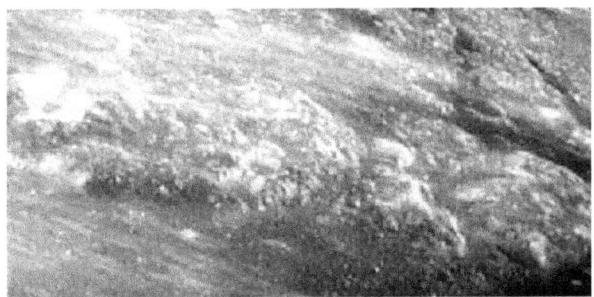

AS15-94-12753
Close-up of Mare Crisium, (Apollo 15)

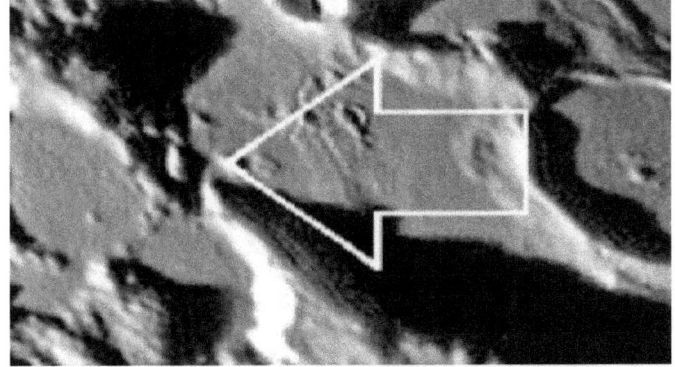

THE BRIDGE
(Compliments of Alex Norman, 'Anacortes Astronomy Club')

This is where Wilkins supposedly confirmed his 2 km wide suspension (or portable) bridge. What we see is the so-called bridge between the Promontories ruined, collapsed, and aged from constant micrometeorite bombardment. With a good smaller telescope, the bridge is often seen by the way the sun hits the mountains and craters in the area.

Of course, no one knows exactly what O'Neill saw because the "bridge" does not exist. However, what do exist are shadows and some highlights. A crater on Promontorium Olivium close to the gap can resemble a natural arch under the right lighting conditions, when seen with south up. Remember, significant changes in the shape and appearance of lunar surface features can occur in as little as a half-hour of watching, so be patient and have

157

fun. Some other views of Mare Crisium can be viewed in Lunar Orbiter photos, IV-061-H2 and IV-191-H3.

18
BRIDGE ON THE FAR SIDE

NAC M113168034R "The Bridge"
Another amazing bit of lunar geology revealed by LROC!

Just when you think you have seen everything, LROC reveals a natural bridge on the Moon! NASA has discovered an amazing

158

lunar bridge on the far side of moon, something Mr George missed. This one just happens to be a real bridge and is visible in the following photos no matter what time of day.

The discovery was made by the Lunar Reconnaissance Orbiter Camera, which sent pictures of the King Crater on the moon's far side to NASA. The bridge is seven metres wide and 20m long and spans a canyon between two and four stories deep. NASA speculates that the formation may have come about following the collapse of a lava tube. Of course, George would have identified it as some kind of alien mining operation.

The LROC confirmed a theory that the moon's surface may hide a vast network of lava tubes – alien tunnels! The discovery of the bridge seems to have further bolstered the theory. Scientists believe that the astonishing formation is the result of ancient lava flows, which left hollow tunnels, which in this case have fallen away and created the bridge like structure, NASA said.

This above photo is one of two natural bridges found within the melt pond region north of King Crater that probably formed by the collapse of the upper cooled surface layer of a subsurface lava tube or chamber [17]. [See also images: M130863593L/R [a, d]; M113168034R [b, c]; NASA, GSFC, and Arizona State University]. The bridge is approximately 7 meters wide on top and perhaps 9 meters on the bottom side, and is a 20-meter walk for an astronaut to cross from one side to the other.

Is this feature truly a bridge? Look closely and you can see a little crescent of light on its floor. That patch of light came from the east, under the bridge. In other lower resolution images, you can see light passing under the bridge from the west. Therefore, there must be a passage.

How did this oddity "bridge" form? Did aliens dig it out and then abandon it? The impact melt deposit on the north rim of Crater King, which is over 15 km across, was emplaced in a matter of minutes as the crater grew to its final configuration. The most likely answer is the surface collapsed into a lava tube leaving a small bridge connecting the sides, while the other sides collapsed into holes. This reminds us of Florida sink-holes that have gobbled up houses and even people.

From the Apollo era and the new LROC images, we know now that lava tubes formed in the ancient past. These images have raised the tantalizing prospects that lava tubes remain intact to this day. In fact, there are possibly many more. The same NAC (Near Angle Camera) image revealed two natural bridges – not just one! Wow! What a great thing to find. We can now design tunnel villages for future lunar colonists and not have to transport digging machines to the Moon, like the aliens foolish did.

To understand lave flow tunnel tubes we have to understand the geological properties of crater development, crater ejecta and lava flow. The impact melt that was thrown out of the crater, pooled on the newly deposited ejecta, allowing its interior to stay molten for a long time. As the local terrain readjusted after the shock of the impact, the substrate of this massive pool of melt jostled to some degree, thus collapsing sections of the upper part of the lava tubes.   Local pressures built up and the melt moved around under a deforming crust. Then the melt was locally pushed up forming the rise and the magma found a path to flow away, leaving a void, which the crusted roof partially collapsed. This is common, for the whole area has pot holes, sink-holes, collapsed lave tube pathways and many other odd sunken spots, ditches and pits [18]. But, strain as we may, we find no alien canals, holes, ditches or transit systems.

There are actually six NAC images in which you can find the bridge under varying lighting: See LROC photos M103725084L, M103732241L, M106088433L, M113168034R, M123785162L, and M123791947L.

19

ALIEN ARTIFACTS ARE THE PITS

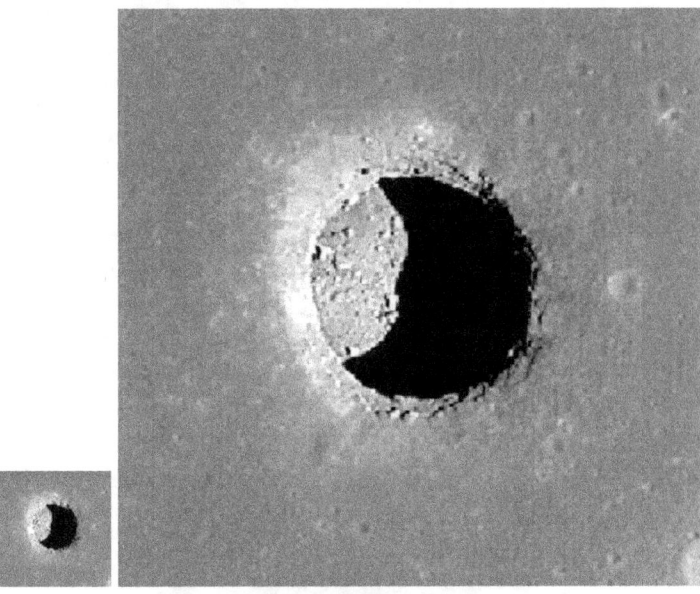

NAC M126710873R (LROC)
"Bullet Hole" or the "Pit" in Mare Tranquillitatis.
[NASA/GSFC/Arizona State University] (LROC)

There are many other related types of lava tube collapses. There are potholes, portholes, sinkholes, collapsed lave tube pathways, and many other odd shaped sunken spots, ditches and pits. There are even what appear to be "bullet holes." The LROC has now collected the most detailed images yet of at least two lunar pits, quite literally giant holes in the moon. Scientists believe these holes are actually skylights that form when the ceiling of a subterranean lava tube collapses, possibly due to a meteorite impact punching its way through. We can be sure that the floor of these "holes" are the tops of the ceiling that collapsed, leaving a large underground mound. Thus, the floor we see is not the floor of the lava tube, but the collapsed remains of the lunar surface.

 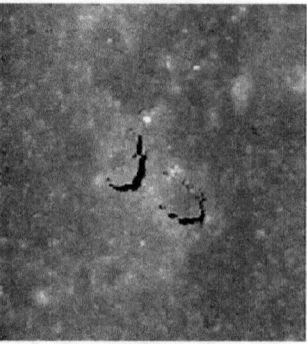

M113168034R   **M123791947L**

The left close-up shows the bridge when the Sun is 42° above the horizon and the right is the same area when the Sun is 80° above the horizon (near noon). M113168034R on the left, and M123791947L on the right, both are measured across at 128 meters.

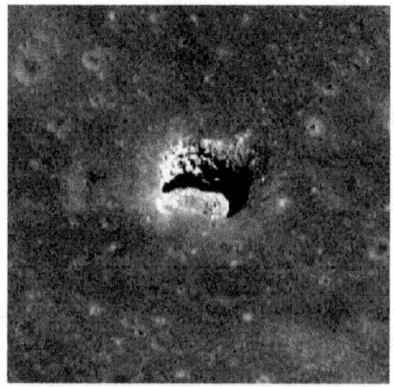

THE MARIUS HILLS PIT

M133207316LE, M122584310LE, and M114328462RE

Image Credit: NASA/Goddard/Arizona State University. (LROC)

The above bullet hole shows a spectacular high Sun view of the Mare Tranquillitatis pit crater revealing boulders on an otherwise smooth floor. Image is 400 meters wide. North is up [19]. See the little alien golf balls littering the floor of the pit?

The Japanese SELENE / Kaguya research team observed one of these skylights, the Marius Hills pit, multiple times. With a diameter of about 213 feet (65 meters) and an estimated depth of 260 to 290 feet (80 to 88 meters), it is a pit big enough to fit the

White House completely inside. The image featured here (middle, above) is the Mare Ingenii pit. This hole is almost twice the size of the one in the Marius Hills and most surprisingly is in an area with relatively few volcanic features [20].

20
LUBINICKY JUNKYARD

AS16-M-2493_med. Bullialdus-Lubinicky Area

The Bullialdus-Lubinicky area is another left-behind alien graveyard site of busted, decayed, and disintegrated Moon moving machines! For Mr Leonard this site offers the finest in used E.T. technology!

The area is located between Mare Cognitum and Mare Nubium, a little South and East of the exact center of the Moon's Near Side. It is exactly South of Cognitum and West of Nubium. It is a medium size crater and somewhat of an old one submerged in the mare surface, and locating it through a telescope can sometimes be an eyesore. Looking for it in NASA photographs can also "crater" the eyesight and after long hours of searching for artifacts can cause a false sense of dyslexia as well as fantastic hallucinations as is demonstrated in "Somebody Else is on the Moon."

164

AS16-124-19901

This shot of AS16-124-19901 is one of the most beautiful landscape shots NASA has taken of this area and it is a favourite public relations photograph of the Moon. The area is full of hills, mounds, dirt piles, crater ejects, pits, gashes, crater rays and all sorts of other morphological shapes that are exciting to the eyes, especially to Mr Leonard and other extra-terrestrial imaginers.

Mr Leonard describes the area as screaming of underground inhabitants - from all the seismic rumbling and general ground disturbances there - and contends that the area is more than (as his associate criticizes) just "Mountainous rubble." He boasts that just two square inches of the photo *can keep one busy for weeks.*"

He said the thirty-seven-mile-wide crater Bullialdus sits in the middle of the southeast quadrant of the Moon and claims that between Bullialdus and Lubinicky E is *"the most fantastic area on the Moon"* riddled with all sorts of *"macroscopic engineered objects"* (p. 40-41).

He continues that the crater with the sun-struck rim (Lubinicky A) has what looks like a shaft of a gear sticking out. We are told to look just below it at the remains of another larger "gear" that apparently has been ripped away, exposing its inner teeth. He estimates that it measures about five miles in diameter and would be a real blast if dropped on Manhattan: *"It would obliterate everything from midtown to the Bowery"* (p.42). Next, we are to note the *"perfect symmetry"* of the underside arc, the *"absolute perfection of the teeth"* in the small gear and the *"four perfectly spaced rows of teeth*" in the larger one. He points out that the shaft *"sticks straight out for at least two miles."*

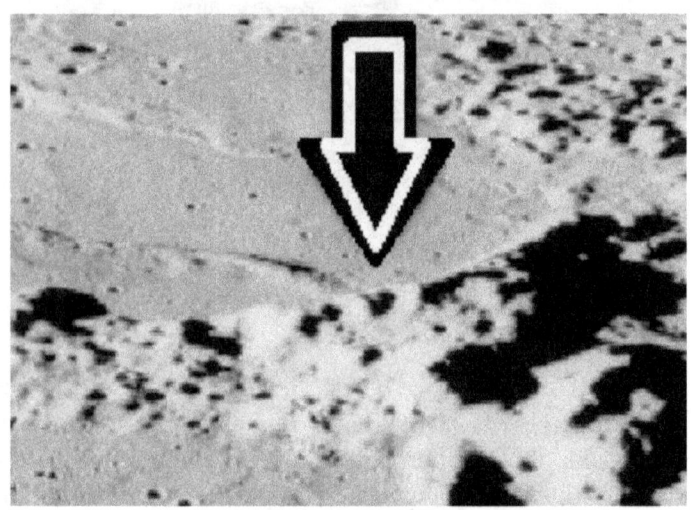

Gear teeth sticking out of rubble?

Notice the crater rim with the smaller meteorite impact holes.

Without showing the reader any photographic enlargements as evidence, he sketches for us some of the engineered objects to illustrate for us what he sees. After enlarging the photo, rather than finding "stitches" the observer will contract stitches or at least a few chuckles. Nowhere are there any artificial machines. There are no parts of equipment or other alien devices. All that can be seen is dirt, rocks, hills, mountains, craters and crater ejecta. Sorry, the alien mystery is unfolding as an alienating scam to distract people away from truth, while making boring space science more fun and exciting!

More gear teeth.   (Unknown)

166

Moon artifact hunters consider this site to contain leftover parts of alien machines. Mr Leonard, (especially) sees HUGE gears and shafts complete with their housing encasements. He said they are artificial constructions and not natural ones, because the sun shadow angle outlines the symmetry of the underside arcs and the perfection of the gear teeth as they cast shadow lines on the mounting plate for the gears, in regular, spaced rows. One is said to be able to see two shafts and their shadows, sticking out over the lunar landscape. The gear housing is at least 3-MILES LONG.

One shaft is at least 2 Miles long. Others may be longer or shorter, depending on the design and usage. As far as other machine parts, there are metal T-Scoops, pipes, busted hoses and - well, the list is endless if you have an endless imagination.

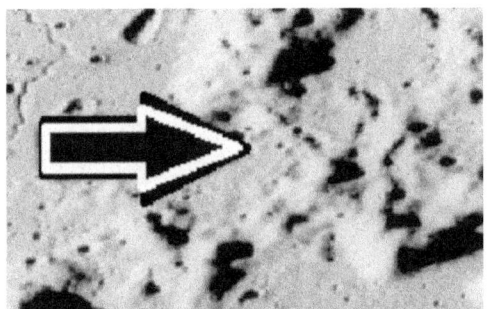

The 3-mile Shaft

AS16-M-2496_med

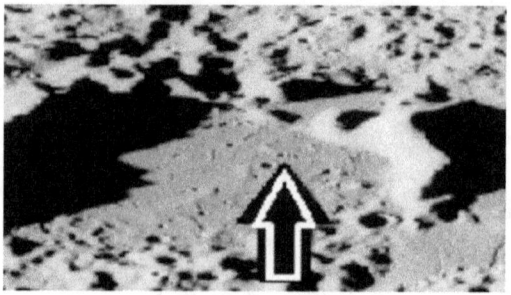

Unknown object

Our alien imaginer said the Lubinicky area is comparable to Crater Tycho in abounding with piles of junk. However, this is not all the junk seen there. There are what appear to be "stitching" or big bans that hold two large chunks of the lunar crust together, as if to hold it from splitting apart. Craters Tycho and Lubinicky are unique spots for these stitching's.

He said, *"These (photos) are some of the best examples of 'stitching' the skin of the Moon in the Bullialdus -Lubinicky area. You cannot fail to notice the precise regularity of the stitches. They are identical in length, identical in distance apart, and identical in width."* Now, how an alien can stitch together loose, dry regolith dirt and hold it together is way beyond science and reason – it is fantastic!

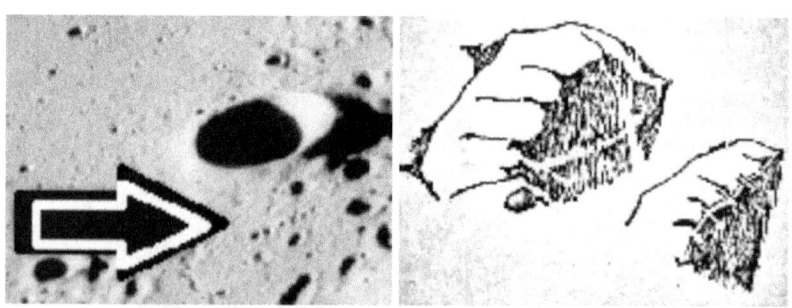

Tycho "stitching's" Lubinicky stitch marks
(Un-located)   A crater chain?

Leonard draws up some sketches to illustrate what he sees, rather than enlarge the photo and show what is really there. After enlarging the photo, rather than finding a "stitch," the observer finds a switch. Nowhere can any artificial machines, parts of equipment or other alien devices be found, so we see why George switched the photo of the stitch with a drawing. All that can be seen in the photo is dirt, rocks, hills, mountains, craters and crater ejecta. All that can be seen in his drawing is everything but dirt, rocks, hills, mountains, craters and crater ejecta.

Examples of "stitching" in the Lubinicky Area
(Cannot locates these)

After establishing the reality of the stitches, ditches, screws, shafts and gears, George sows together another tapestry of industriousness, explaining why aliens were stitching up the lunar surface. He finalizes his polyester yarn with the moon being an interstellar spaceship that had received major damages for some reason and was (is?) in the process of repair. This is at best a flying carpet story if you think twice about it or can think at all.

169

21
KING OF THE FARCE SIDE

M119062083MC
King Crater on the far side of the Moon - LROC

George Leonard seems to be the KING of the dark side of the moon studies. King Crater is another of his "fantastic" locations. If two inches of Lubinicky regolith would keep you busy for weeks, this King of the farce side will have you buried in it for years to come.

King is a prominent lunar impact crater that is located on the far side of the moon, and is not viewable directly from Earth. It forms a pair with Ibn Firnas, which is only slightly larger and is attached to the northeast rim of King. To the northwest is the crater Lobachevski, and Guyot is located an equal distance to the north and northwest.

170

King Crater is 77 km in diameter and has what appears as an unusual claw shaped central peak or what looks like a fat tuning fork.

This is not an alien amphitheatre, but is the rebound material going up and collapsing to the North. There are landslides in the walls and impact melt ponds on the floor with some volcano like domes with vent-like craters in their tops running in a line along the northeast floor. The south wall of King has collapsed inward onto the floor and external to the rim on the ejecta blanket to the east (right) it has a flow like appearance. There is just so much to see and describe in this King of the far side of the moon crater.

Leonard also presents some interesting alien machines digging about King Crater and the surrounding areas, and something coming out of a small crater in the unnamed highlands. He said, this mysterious object was *"coming up out of it."* This was the largest "spraying" ejecta stream he had ever seen other than Yellowstone's Old Faithful. He finds in the King's area what he calls digging machines or "X-Drones" and other industrial items. Now, we will have to take his word for most of what follows, for all the looking in the world is not going to find one alien iota.

There are other reasons why we will never find what George sees. The first reason is that most people are not brain surgeons and cannot tell what is happening in his imagination. Secondly, many of the "artifacts" cannot be relocated because he used the public affairs office press release renumbered photos rather than the NASA photo

M103739394LE
Deceleration lobes in King crater ejecta. LROC Narrow Angle
Camera (NAC) observation, LRO orbit 453, August 1, 2009.
Lighting incidence angle 61°, field of view approximately 1,200
meters across.)

code numbers. The number system was there for reference and it is
obvious why he did not use the code numbers: He was too lazy and
cheap to make a trip to the NASA archives, buy the negatives and
run detailed copies off from the original masters. This is the reason
for the blurriness of the pictures, which by the way lends support
for the blurry obscured alien evidence. One would have to search
for weeks to relocate most of the little craters. Some of these
craters are located, but most are impossible to find unless a cross-
reference is located.

As far as alien artifacts are concerned, just looking at the
quality of his photo plates begs the question of stretching the truth
or rather "stitching" truth together out of mental phantasms. The
objects are mostly blurry and the ones that are clearly defined look
not much different from millions of other so-called "alien artifact"
objects scattered over the moon.

172

A Constellation Program Region of Interest near the northeast edge of the unusually large melt pond adjacent to the lunar far side crater King. The boundary between the dark, coherent impact melt rock at the lower left of the image and the bright, pulverized ejecta blanket to the upper right is clearly visible in the floor of a smaller crater that formed at the boundary between these two units. Image width is 1.3 km, pixel width is 1.29 m

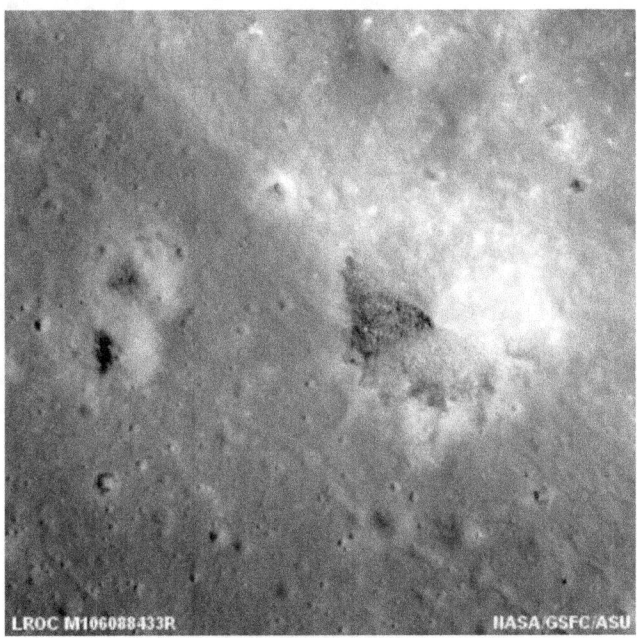

Subset of NAC frame M106088433R

M115529715RE

A fault scarp separates two zones of impact melt within the King Crater (5.0°N, 120.5°E) north interior wall. LROC Narrow Angle Camera (NAC) observation, LRO orbit 2159, December 15, 2009; solar illumination incidence 75° from the east, field of view approximately 900 meters (north is up). [NASA/GSFC/Arizona State University].

22

REGOLITHIC ROVERS

NASA is working on what they call a Spray-puff, smoke blowing EXO-robotic recreational vehicle or colossal size roving regolith mineral sampler. Leonard anticipated the invention decades before by speculating that aliens had invented them long before man ever even had the wagon, not to mention the steam engine to pull it. In the following photo of Crater King, he locates such devices to prove his point.

AS16-120-19228

PLATE-12. Circles indicate several small craters in the process of being worked with marking crosses on their lips and spraying drones inside. Also X-drones and puff- like orbs circled in this shot of area near King Crater.

In the above Apollo 16 photos, Leonard mentions the *"technological glory of the Moon occupants"* left behind. This junk apparently is in use and lies between King Crater and an unnamed crater with a "pond effect" bottom. *"There is a long ridge separating King Crater from the ... crater with a smooth pond in it.*" (Many craters do have flat bottoms according to lunar morphological process).

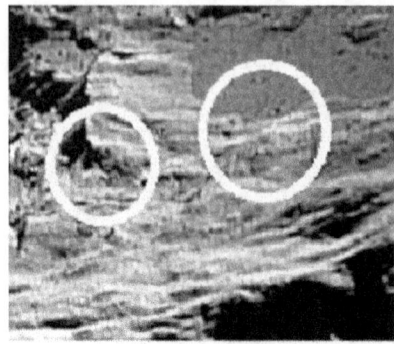 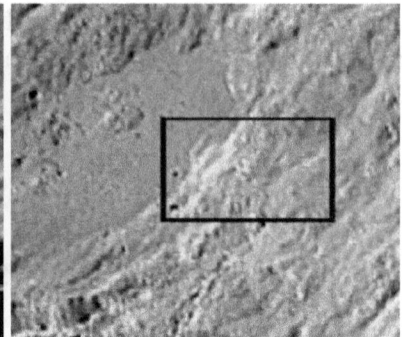

72-H-834 (AS16-120-19265)   AS16-120-19226
X-drone with no spraying.   Another X in King Crater

He said, in one of the areas of King "*Something was coming up out of it*" outside of the rim area. This was the most dramatic "*spraying effect*" of all the spay-jobs he would ever find. Mysteriously, this active "spray" in photo Frame No. 834 does not exist in the later photo No. 839 taken two days before. He said this was his amazing proof that an artificial spraying was taking place. Now you see it and now you do not. What is more amazing is that No. 839 (AS16-120-19228) is a more medium shot, further away, and if you look at it closely, you cannot make hide-nor-hair out of it, let alone make out any alien "hides" or E.T. "hairs."

Nevertheless, Leonard has found more astro-goodies to support his quackery! With super-human eyesight, he locates an X-drone in the earlier 839 picture and said it must have been "*working in exactly the same spot from which the spray would emanate in the 834 picture taken two days before.*" With even more super-human vocal cords, he squawks further of X-drones digging for minerals and such, but speculates other reasons for excavation activity. The aliens need locations for "landing berths," "archaeology," "recreation" and places for "competition" games.

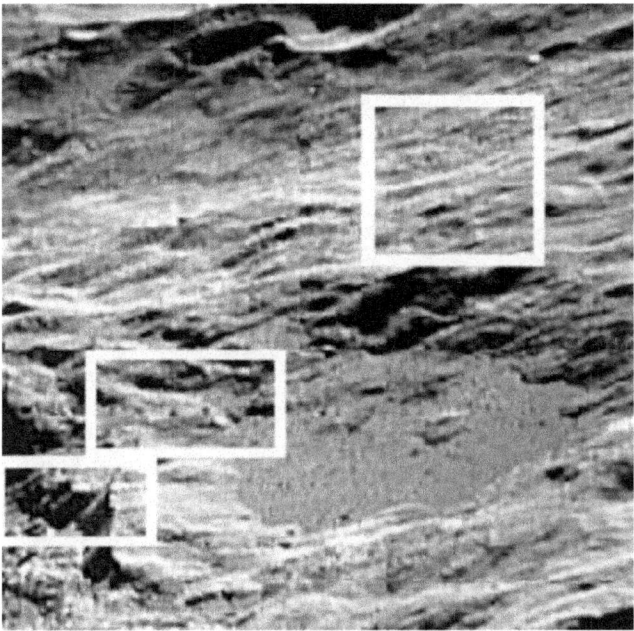

AS16-120-19226
Top box shows the "sprayer" unit.

At the base of the ridge (in the above picture), he said are *"some old friends and some new oddities"* that *"began to jib together, and to click."* Yet, after some detailed observations of the photograph, we find rather George has conned his new friends by fantastic interpretations of old oddities. One little jibbing object looks like a large *"oversized cannon"* that could be used as some kind of recreational game device, a mechanical digger or *"some functional rig in a complex series of steps undertaken by a complex culture."*

No matter, after showing the reader a pathetic drawing of a blob with a suction hose for a nose, he begs his readers to come up with their own ideas about what it is. Since he cannot tell a moon mountain from a molehill, he suggests that we supply a label, since he does not have enough evidence for an idea of what to call it or what its function may be.

What a pleasure it is to suggest a few things for Mr Leonard. It is an astro-biogenetic dinosaur size Anteater! Yes, take a new look. It looks like a huge colossal size balloon bag. Could it be a

177

747 size alien colonic bag? Could it be a gigantic trampoline, part of a bagpipe or a giant exo-kazoo? What if it is only a brightly sun lit funny shaped rock? Maybe we will never know, because the photo is too grainy to tell.

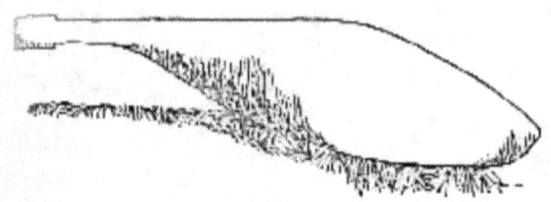

Leonard's attempt to draw an alien device in Crater King. Looks like a huge "cannon." (p.80)

23
X-DRONES OF CRATER KING

Here are the drones drawn by Leonard. The arrows point to their locations according to his specifications.

The "X-Drones" of King Crater

Leonard did not publish very good versions of the NASA Apollo 16 King Crater photos. There is more than one King Crater photograph. I picked the best available from all the frames. Since he also did not enlarge the areas I thought it necessary to show my readers the big "X" drones, where the X stands for an "X" out"! Whatever he saw in the photographs are not there now. Here are some extreme close-up enlargements of the three locations.

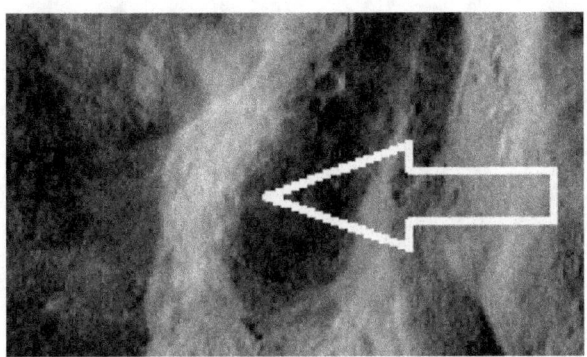

AS16-M-0890_MED -- CLOSEUP -1

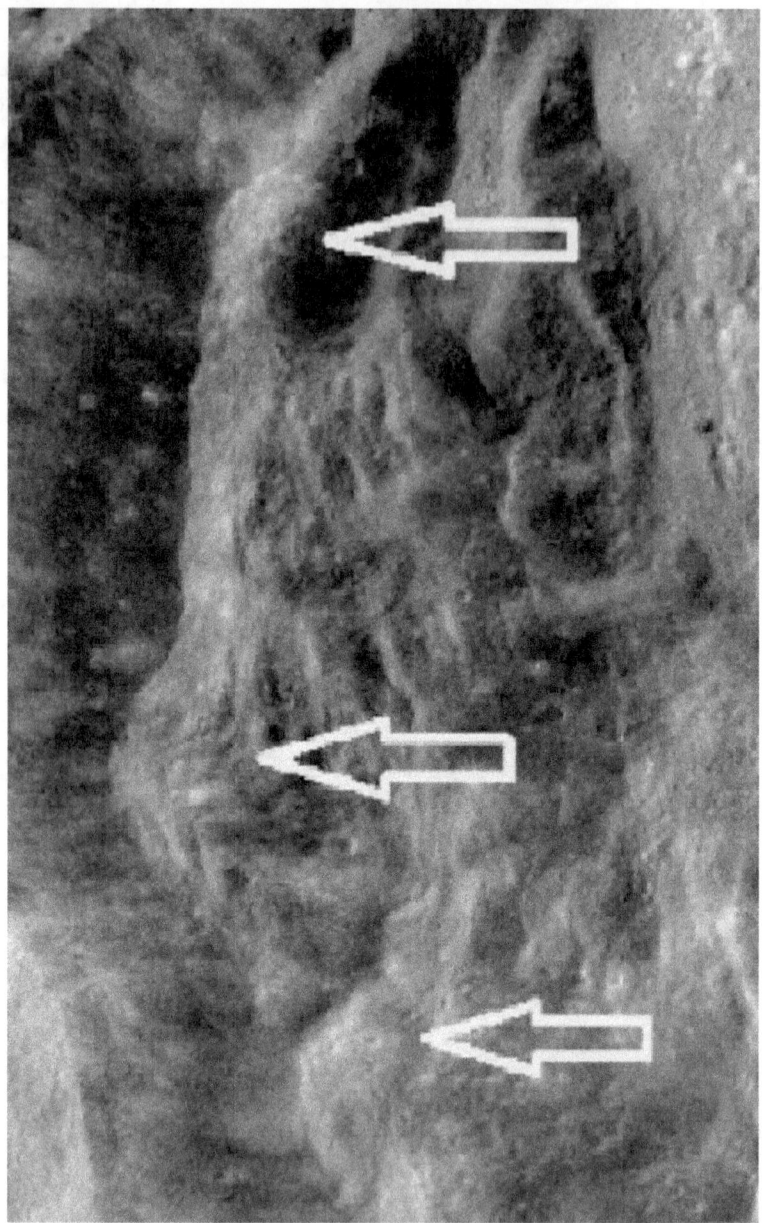

AS16-M-0359_med
Location of the X-Drones

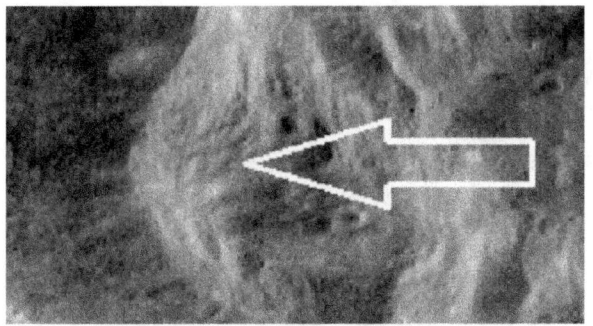

AS16-M-0890_MED -- CLOSEUP - 2

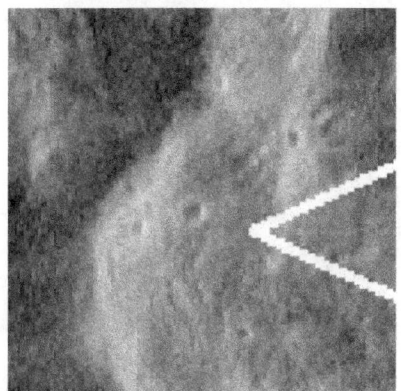

AS16-M-0890_MED -- CLOSEUP – 3
Nothing!

SO we have evidence of crater sprays, ridges being knocked down, huge X-Drones slaving away for their masters, grey lava sludge smoothing over lunar surface , pounding effects stirring up dust, little puffy orbs and nipple like extrusions - what more could be found? The possibilities are endless depending on the excessiveness of your imagination.

24
DINOSAUR CRATER

LO-I-136-H3
Unnamed Crater on far side of Moon. Nickname:
"Icon or Dinosaur Crater"

The DRONES are radical in *"GOUGING OUT A MOUNTAIN"* in search for minerals, said George! These "X-Drones" are gobbling up and spewing out lunar dust. Leonard suggested (in plate 8) that drones are *"raising dust on the rim of King Crater"* and in other locations, such as the above. Furthermore, he mentions there are many types of regolith rovers

182

and other machines, such as Big-rigs, T-scoops and super rigs in various places on the Moon.

George directs us to the far side of the Moon for further proof to a location north of Crater Tsiolkovsky. After consulting the surrounding area other oddities appeared (See Leonard, plate 4). Leonard suggests that super rigs are working an area at the base of the walls of an octagonal double crater (pp. 49, 50 and 51).

The photo in Leonard's plate-4 (NASA Photo 66-H-1293) is of an unnamed double crater. It lies directly north of Tsiolkovsky close to the Craters Love and Prager, at -5.40 degrees Latitude, 129.33 degrees Longitude. This fascinating double crater has a gently domed floor, crisscrossed with linear depressions [about 1 kilometer wide] that give it the appearance of a turtle's back and crudely polygonal fracture patterns occur, formed by upward movement of plastic or liquid material. The doming of the floor of this crater could be the product of local magmatic activity or of slow isostatic rebound. Both alternatives point to probable activity of internal origin in the lunar crust [21].

At the base of the rim in the larger crater at about two o'clock, according to Leonard, is an object, *"which is too wispy to show up well when reproduced in this (his) book"* (p.51). He said the wispiness is due to its structure, whereas it is actually due to the bad quality of the photo enlargement. He illustrates this object in the following sketches (p. 52).

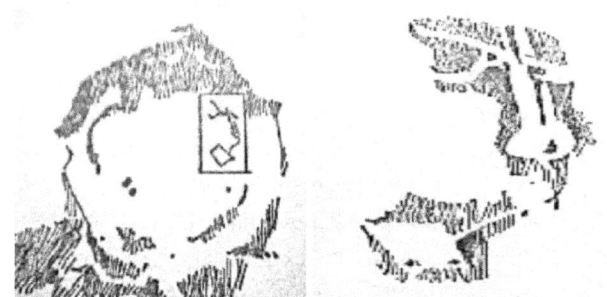

Drawings of crater Big Rig "scoop"

183

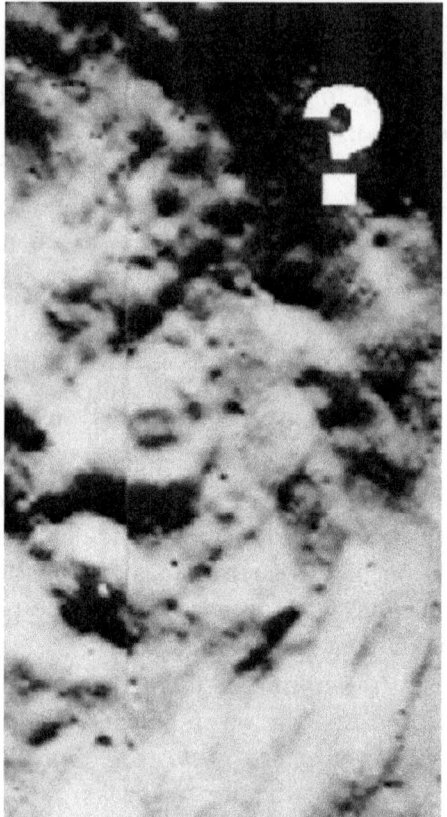

Enlargement of LO-I-136-H3

He continues: "*The object pivots at the junction – just where you would expect it to pivot. Its two main struts rising from the ground (the wispy elements) are very straight and parallel. There appears to be a filament of some sort, which raises and lowers the horizontal piece leading to the scoop. There is a long thin device that runs from the base of the object down the hill toward the center of the crater, ending with an oblong plate, which is found elsewhere on the Moon and is perhaps connecting to a power source*" (p. 52). He goes on to explain the large equipment, the need for resources, minerals, and alien habitat.

Is this really in Crater Pasteur D? Let us look closer at this area and see what we can uncover. Notice in this enlargement that the eye can strain until it bursts before it sees any super rig. Super-size your eye balls, because no normal eyesight is adequate. The following cropped picture is the exact location George is sketching.

184

We will not waist our time looking at the higher resolution LROC pictures, since this one is sufficient to prove the nonsense.

A SUPER-SIZE DINOSAUR

What did George overlook that might be interesting? With little imagination one can carve out of the distorted crater floor and debris geometrical black dotted rectilinear bars, some connected dots (red triangle), a "big-rig" (red square), and three rows of black dots (small red square). There is also a statue of an "eagle" (red oval) and what looks like an artificially "sculpted" super-size dinosaur (red arrow)!

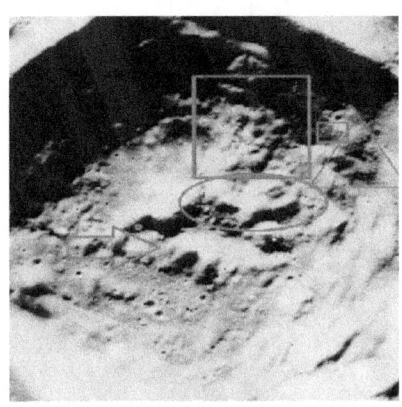

Locations of objects in red.   Dinosaur icon mound

Carved monument of an Eagle

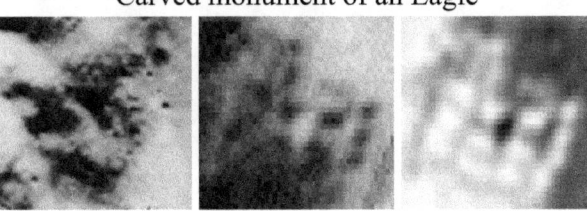

Geometric constructions with bulbs, dots, and "round appendages" along a set of (geometric) ridges. (Mid) 12 Dots pattern. (Right) Negative image of 12 dots.

Leonard's speculation of aliens needing recreation suggests that they busied themselves on their off time studying social science, art, and the history of man. It looks like this might be the case for dinosaurs, eagles, and other animals do not live on the moon and never have. The "occupants" must have manufactured iconographic versions of earth creatures for study and pleasure. Maybe they were toys. It is possible that these so-called icons represent leisure prevailing over labor and that the moon occupants were free to pursue games, recreation, and scholastics.

25
CLOUDS OVER LOBACHOVSKY

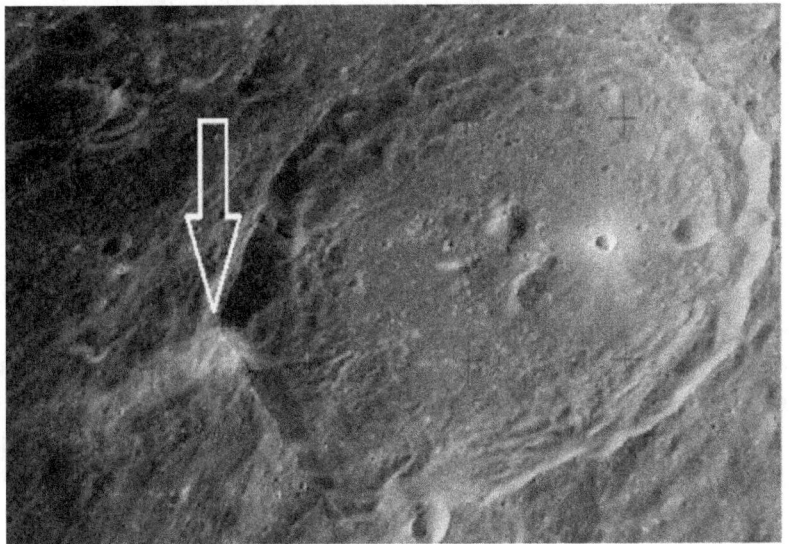

AS16-M-1322

Leonard's 72-H-1113 photograph on page 141 is of Lobachevski
Crater; a crater NW of King Crater on the far side of the Moon a 10
degrees North by 112 degrees West.

Mysterious white ectoplasm energy clouds are now the topic
with cratered scientists such as Mr Leonard. He said Lobachevski
is mysteriously spewing out some "intelligent" energy cloud over
its rim, as can be seen in the photograph above. As to whether it is
a smart cloud, a gas spray, an ejecta RAY or some low level
nimbostratus or stratocumulus cloud, is hard for him to tell. From

187

his reproduction of the photo, it could be some luminescent gas spraying out that brightens by the sun's rays hitting it or a photographic processing glitch. It might even be the smoke spewing out from an alien chimney on a cold lunar night. It could be whatever you want it to be if you are writing a science fiction book rather than a science fact paper. It could also be exactly what NASA said it is a crater ray made from a highly reflective regolith material.

Leonard insists it is not a "ray" like those of Crater Tycho and Kepler, but must be some form of *"pure energy moving over the crater rim towards the center of the crater."* He further said, *"The band of light is not a ray, it is not a patch of reflected light on the ground. This band of light is like no other light one usually sees on the Moon. It maintains its integrity as a light even inside the rim of the crater, which is in shadow, (because) the topography beneath the band 'cloud' shows through."*

Our crater expert then complains that it is still a complete mystery to him as to what it really is. This is no wonder to those familiar with NASA photos and the different resolution qualities. Poor quality photo enlargements cause distorted images and mistaken interpretations. If Mr Leonard had used higher quality images, he would have clearly observed that the rays were the result of fragmented ejecta material.

Crater rays appear when ejecta are made of material with different reflectivity or thermal properties, usually having a higher albedo than the surrounding surface. More rarely an impact will excavate low albedo material, for example basaltic-lava deposits on the lunar maria. Craters on the near side with pronounced ray systems are Aristarchus, Copernicus, Kepler, Proclus, and Tycho. Similar ray systems also occur on the far side of the Moon, such as the rays radiating from the craters Giordano Bruno and Ohm [22].

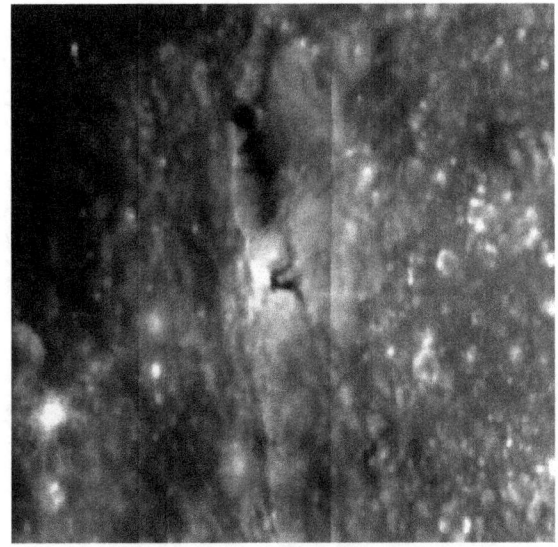

Lobachevski Crater photographed by Clementin

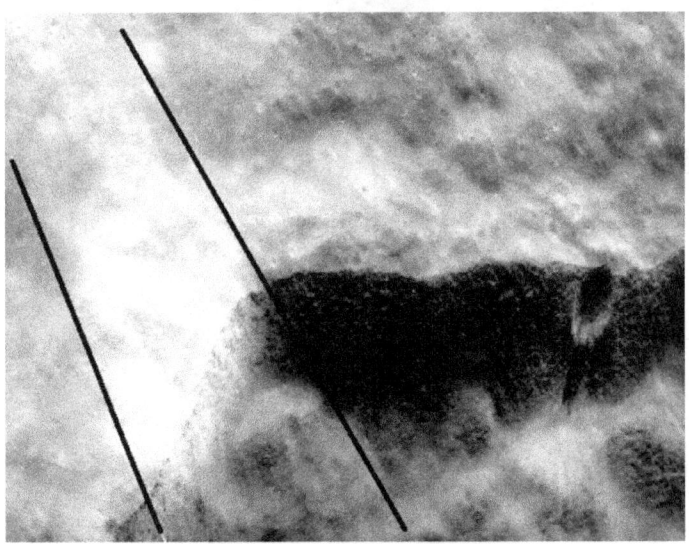

AS16-P-5029_FULL_MED

To cover up the true nature of Lobachevski's ejecta "ray", George took the photo (72-H-1113) and published it without offering the higher resolution panorama images (AS16-P-4849. See also 4851 to 5029).

189

The following Apollo 16 photo is an example of higher quality resolution pictures. Here we can clearly see the ray as part of the lunar surface and crater rim, and not some cloud floating above.

These enhancements are sufficient in demonstrating that there is no mysterious exoplasmic intelligent energy bleeding over the crater rim. When observed in high resolution, it reveals exactly what NASA said it is. It is simply a crater ray just like those of Tycho and Kepler.

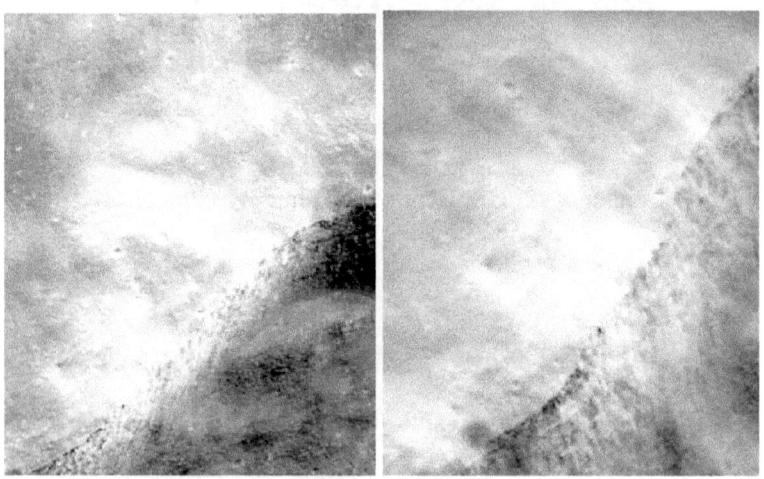

AS16-P-5022_full_med   AS16-P-5029_full_med

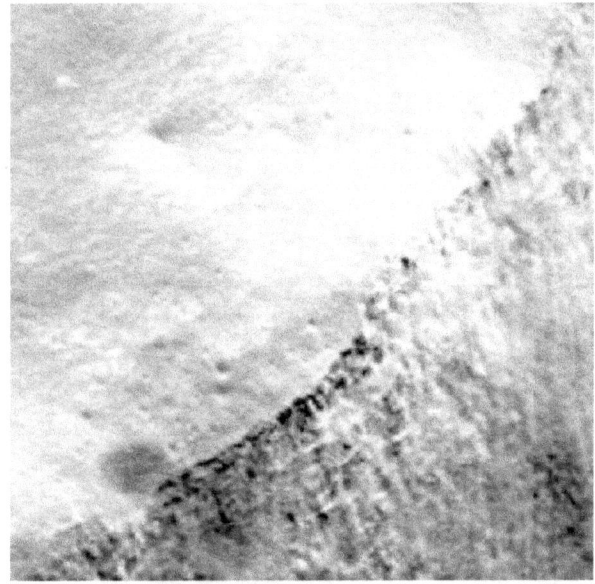

Photo clearly shows higher albedo material as the composition of the crater anomaly.

26
SUPER-SIZE-ME RIGS OF PASTEUR -D

Giant super rig machines are on the moon, said Leonard. In plate 6 of his book, George shows us a photo of a crater with a super rig machine taken by Apollo 14 (See 71-H-781 = AS14-70-9686), which he said is sitting inside. He does not name the crater, but the lunar maps name it Pasteur D. It is located at 9.0 degrees S, 109 degrees E on the lunar far side. It sits on the northeast corner of Crater Pasteur and is viewable in the following frames: LO-II-196-M, AS12-56B-8308, AS12-56C-8308, and AS12-56 D-8308.

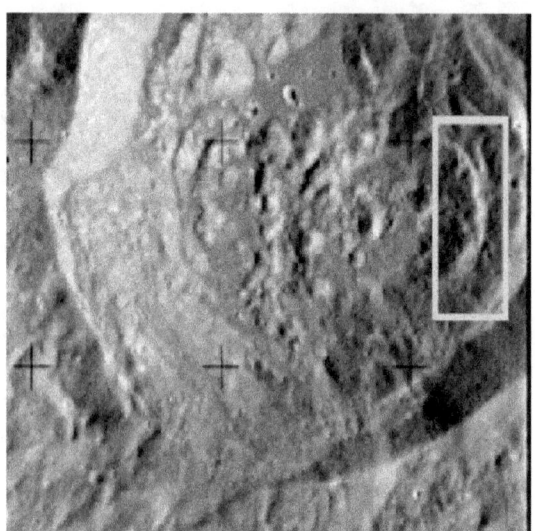

71-H-781 (AS14-70-9686)

George said this is the most remarkable photo ever taken by the astronauts during the Apollo 14 mission. It is the clearest picture of a super mechanical rig on the moon available. He calls this "super-rig 1971" an almost duplicate of his 1966 rig.

It supposedly sits on a terraced, inside rim (See Drawing A) of an unnamed crater (now called Pasteur D) on the far side of the Moon. It stands up straight and is made of filigreed metal for strength and lightness, thus casting no observable shadow. A "cord" runs from its base (B) down the side of the crater (C). On the right of the terrace two other rigs are working. They are of the same design as the first one mentioned and have two pieces

192

working from a fulcrum. Cords run from their bases. They have made an even cut (D) straight down into the terrace. The cut notch is straight as a die. Something also stretches across the gap. As to the size of these objects, George arrives at the rough estimate of one-half miles for the rig, and at least three miles high for the chunk of ground from crater floor to where the rigs are perched (p. 54-55).

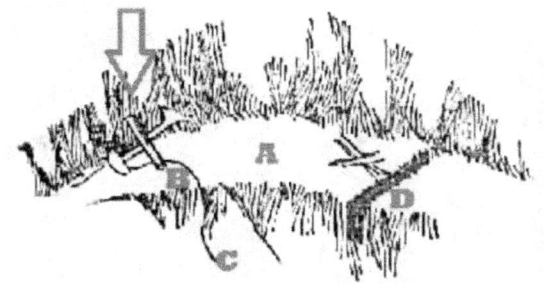

Leonard's Drawing (A) of "super-rigs." (p. 55)

Is this the single most remarkable picture ever taken? There are hundreds of other remarkable Apollo 14 photographs and thousands of other Apollo mission photos. This one photo is only remarkable to George for a specific reason and holds no value to the conservative selenologist. It is no more "remarkable" than the thousands of other Apollo photos. In fact, it is one of the least interesting of the Apollo collection.

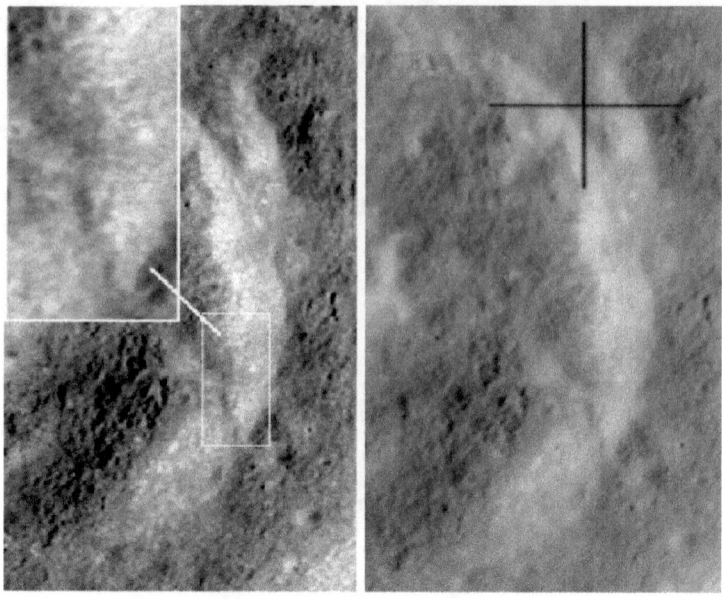

AS14-70-9688   **AS14-70-9689**

Crater Pasteur D PLATE 6. Leonard's "super-rig." Notice no "super-rig" or machines in frames 9688 and 9689 enlargements.

Is a super-size-me rig squatting on the inner side of the crater? We cannot tell from these frames no matter what enhancements we make. It just looks like a collapsed crater rim with heavy sun highlight glistening off the top edge. Nevertheless, to make sure we are not passing up a grand opportunity in finding a priceless antique, alien spacecraft or mechanical digger, we must look closer. None of the Lunar Orbiter or Apollo photos will help. To see what is there, we must turn to the new LRO photos to obtain higher resolution pictures. Notice all that is visible are rocks, boulders, dirt, regolith and crater debris.

The Lunar Reconnaissance Orbiter camera reveals even less about super-rigs and more details about natural lunar surface processes. In no location do we find any machines, diggers, dirt flingers, terrace cutters or dump trucks.

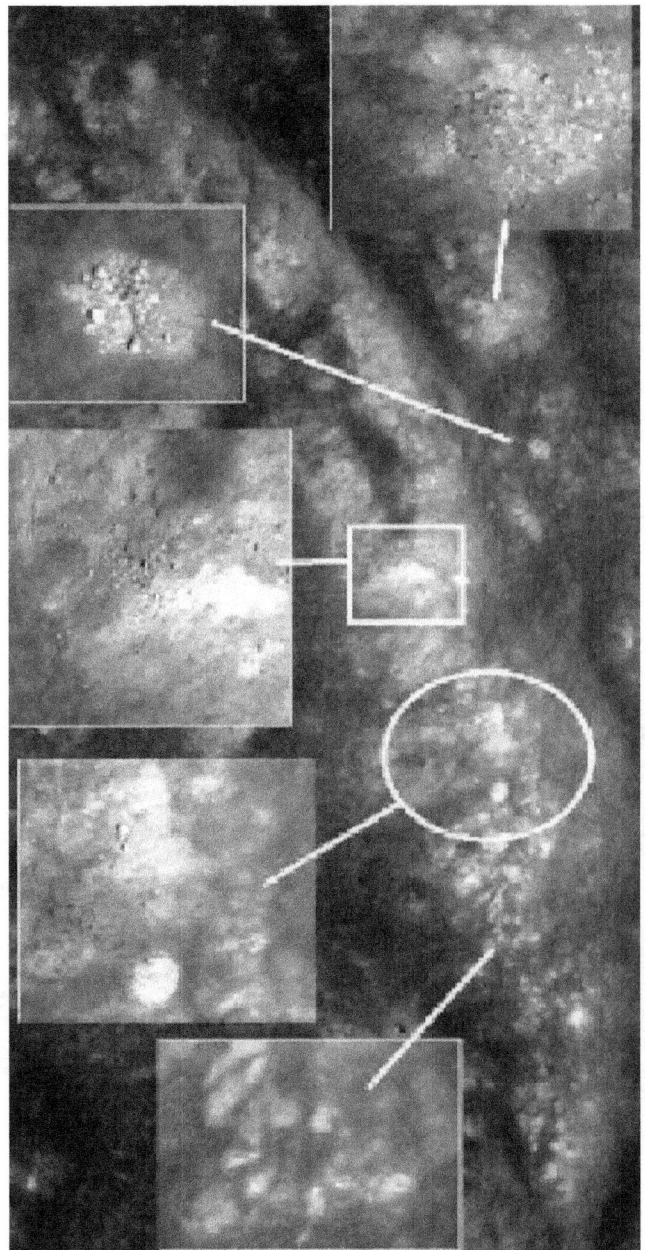

LROC frame M1100495581L
Enlargement of Pasteur D crater rim and "Big-Rig" area.

SUPER TREADS AND LUNAR LADDERS

Next we are told to see some "tread marks" in the following Apollo 8 picture of Crater Doppler, located 12 degrees S, 162 degrees W. on the far side of the Moon. If it is not a super rig, it is "super-treads."

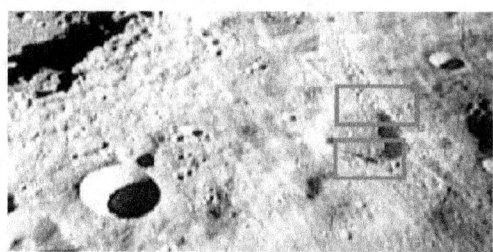

69-H-8 (AS8-13-2244)

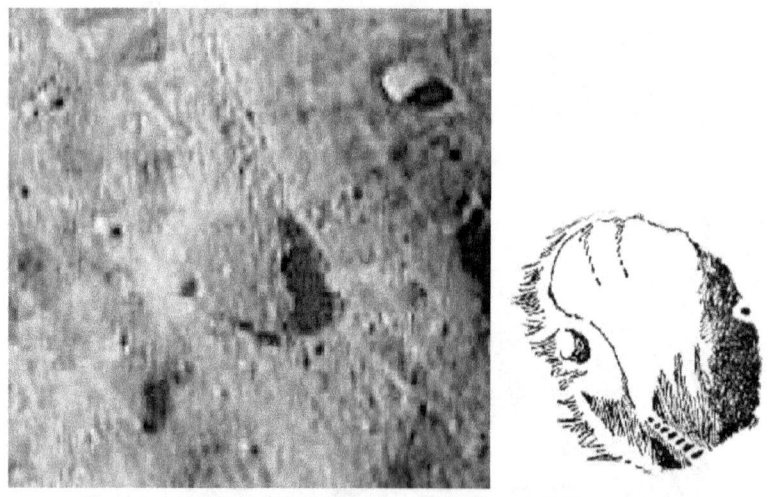

Close-up   Page 175
Drawing of crater "rope-ladder"

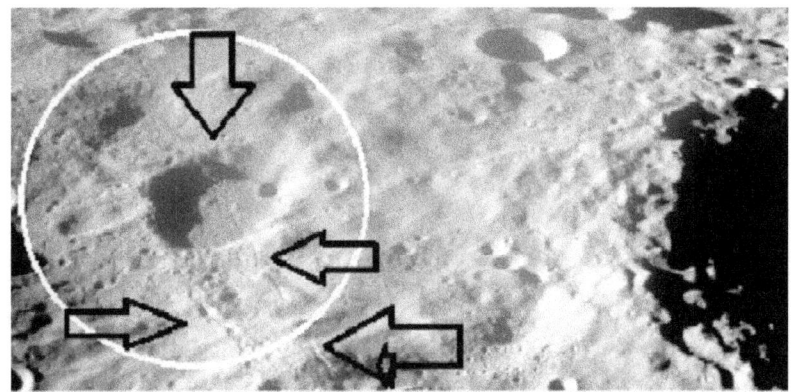

AS17-151-23115
Crater Doppler area

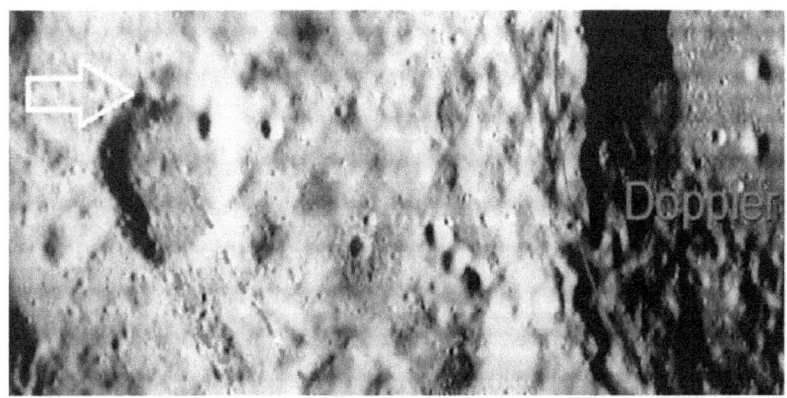

LO-I-036-H3

Leonard goes through a lengthy diatribe explaining some kind of construction was going on here. He mentions, there are *"tread marks on (the) rim and (a) rope ladder extending from (the) rim to (the) floor... visible on the far side crater"* photographed by Apollo 8."

He said, *"In it [this far side crater] is an almost obliterated crater with many parallel markings that run through it with one marking running in the air from the rim and into its bottom."* It appears to be "an enormous rope ladder" or "tread from a very large vehicle." He concludes from a rough measurement that it is four miles long. He draws what he believes he sees and would have us believe and see what he wants us to see, because the picture resolution is too weak to detail it. Do you see a "ladder" in any of these photos?

28
WHAT LEONARD MISSED

NUMERALS   LO-II-037-H1   LETTERS
The number "4"   "Exclamation Point"?

Giant "C" Clamp   LO-II-108-H1   "Rope, Cable"?

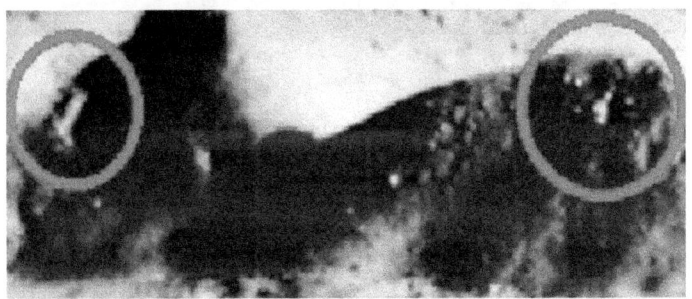

Sledge Hammer?   LO-II-162-H3
Weird device, "Clamp"?

199

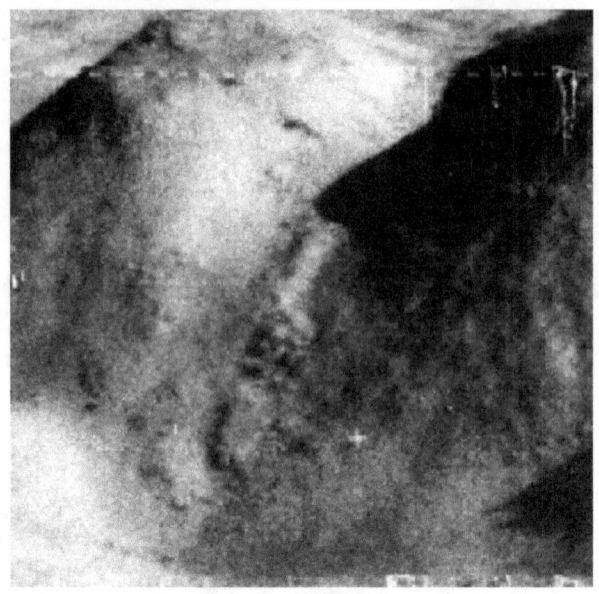

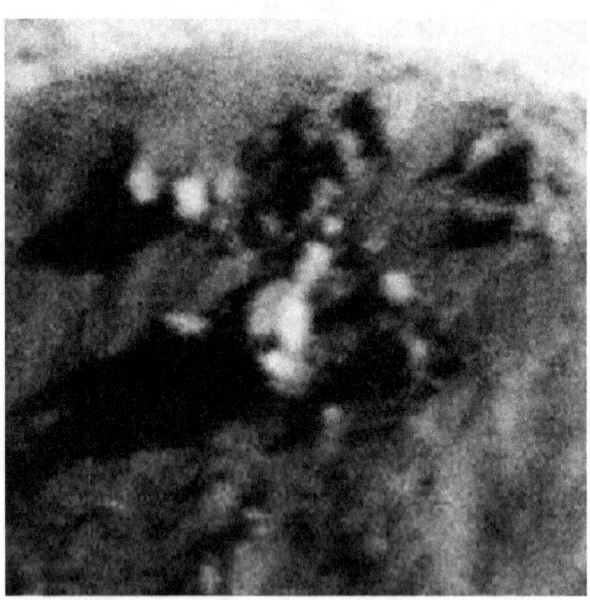

LO-II-162-H3

Just dirt and rocks! (LO-IRP enhancement: "MoonView.com")

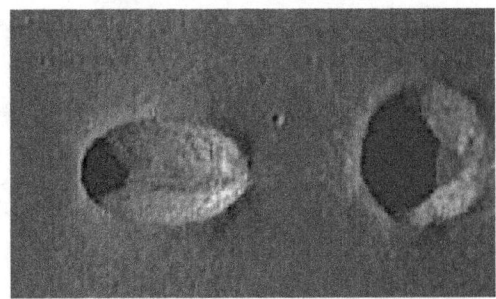

"Coffee Bean" Crater
AS10-33-4906

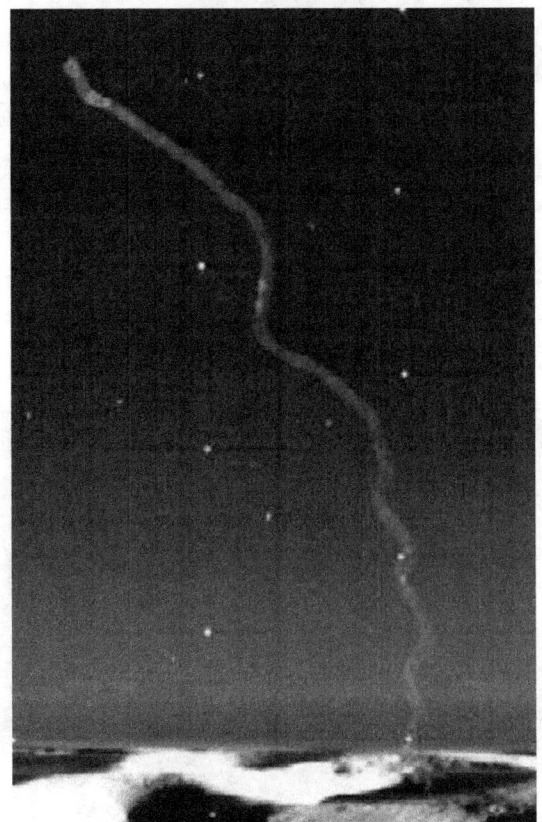

Smoke Spiral Crater
LO-III-213-H1-[2]

201

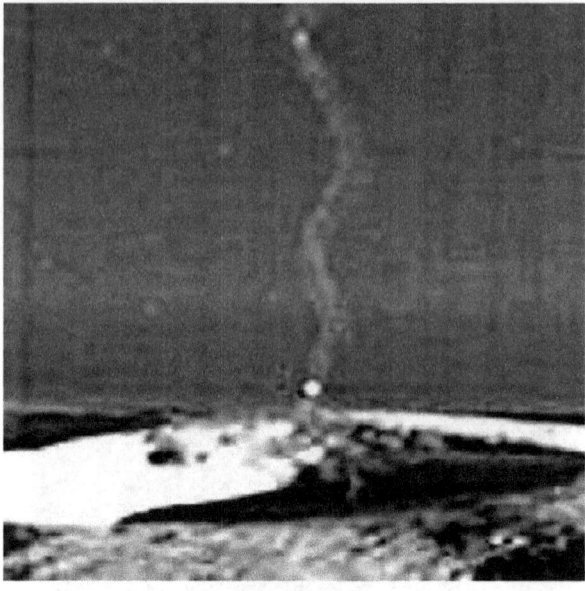

(Close-up)

While enlarged, one can see that the smoke spiral is continuously the same "width". This tells us that it is a "traveling emulsion" blob processing error.

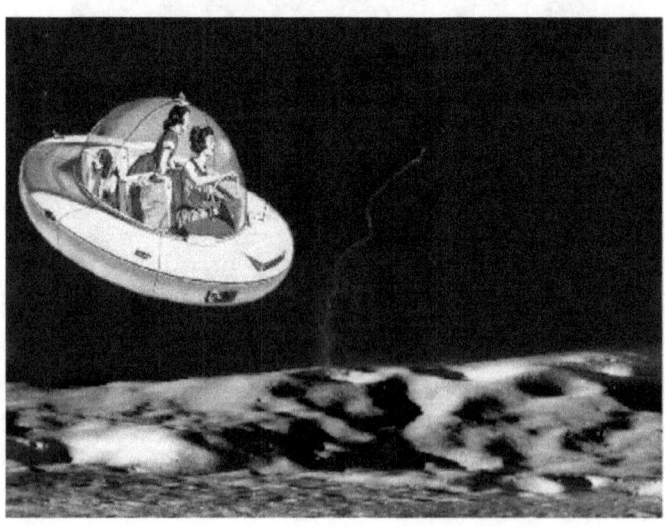

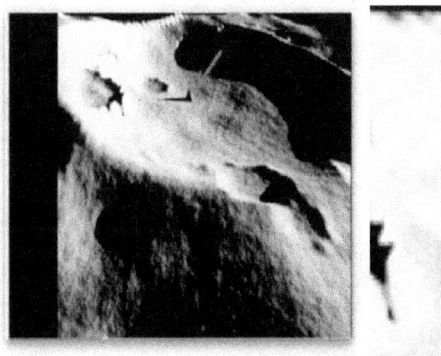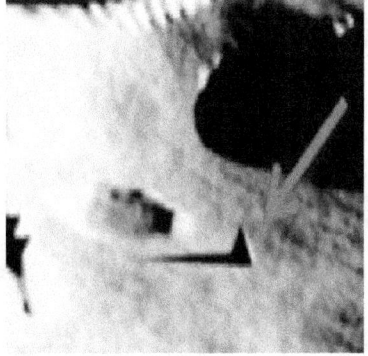

Snap_2011.07.13_11h25m39s_005
PYRAMID IN CRATER (Hubble telescope, 2008)

NASA uses Hubble to look for resources on the Moon too. Therefore, it is possible that it could have taken this picture. Every mention of it on the internet said this picture was "leaked" out from the Hubble classified archives. Of course, no one has traced it to the Hubble. Some say it does not exist in the Hubble archives. It is suspect as a fake because of the perfect symmetry of the pyramid shape [23].

Here is a nice geometric and hieroglyphic shaped object sitting in a plains area. It looks artificial from a distance, but up close shows the same extreme highlight and shadow contrasts as all other similar types.

AS15-M-1554_MED

AS15-M-1554_MED
(Enlargement)

NEED BETTER GRAPHICS EXPERTS!

Further studies on Lobachevski Crater found other "things" or rather created things by Moon artifact hunters. Leonard missed one of the best alien "monuments" for consideration by not referring to the higher resolution images. If he has done so, he might have gotten to this "tower" first.

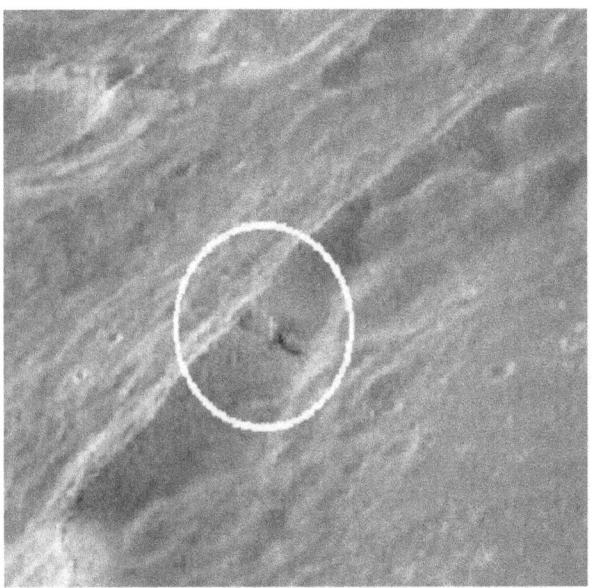

AS16-121-19407
Mysterious crater "tower"

Miss X's TOWER

After lying dormant for decades, this picture of Lobachevski Crater was presented to a graphics analyst, Miss X (we shall call her, "Miss X") at 'I Wonder Productions' for enhancement and further analysis. Her work quickly underscored ("distorted") the facts and therefore, incorrectly showed some bizarre anomalies in the photograph. Furthermore, because of her special graphics enhancement ability, she was able to show more than one. Nevertheless, for brevity, we will deal with the most prominent so-called object - an object that is actually not there!

Miss X thinks she exposed what appears to be a tiny tower in this Apollo-16 photograph (AS16-121-19407). As the "story" goes, she made it clear that her "multi-stage enhancement process" takes the image to its most critical viewing size "without distorting" the object in question.

Below is her first enhancement of the NASA Apollo 16 photograph. It is a shame as well as a sham that such a so-called graphics expert does not multi-stage "enhance" the available higher resolution pictures.

Her excessive computer enhancement process of a lower resolution image actually creates an optical illusion of a distinct object through pixel distortion of a small highlighted rim area. In the process of excessive enhancement this highlight becomes pixilated into an over blown rectilinear "tower".

The following are a series of improperly (excessively) enhanced frames of 19407. The following picture demonstrates that any lower resolution photo can be distorted to shows unusual images, illusions and fake objects. The excessive enhancement of any photo can easily create objects that are not there.

In stage 1, frame 19407 is enlarged. Stage 2 adds enhancement and sharpening. Picture No.2 above shows the actual impact crater with sun highlight on the right side of the rim. No. 3 is stage-2 of enhancement process.

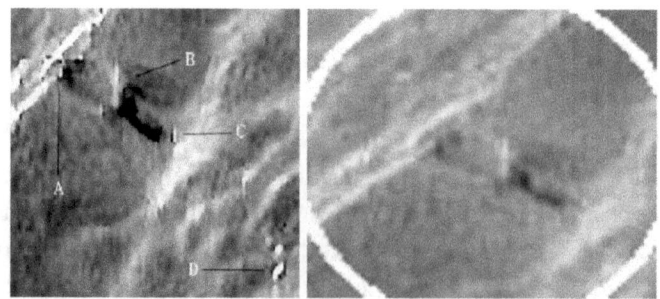

AS16-121-19407

LOBACHOVSHY TOWER

(2.) AS16-P-5024 Enlarged

Notice the rim is highlighted [1] and there is no mysterious object "tower" [2].

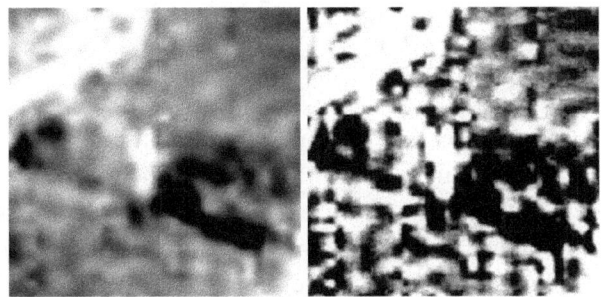

(3.) Ms. X's technique replicated - stage-1 of 19407
(4.) Stage -4 Enhancements
(5.) Enhancement doubled = extreme distortion!

The sharpening is doubled to 210%; radius is 220% and zero thresholds. No.4 is an exaggeration, all enhancements again doubled. No. 5 is all the above doubled again. What we now have is a bright white tower or rocket shaped object, all created out of false enhancement processes. The picture below is the actual craterlet impact site enlarged to maximum size before distortion begins.

Her false analysis is beyond remarkable, in that it consequently creates other strange and unusual 'objects' or anomalies on the rim of Lobachevski Crater. She describes object B as a spectacular 'spire' soaring perhaps hundreds of meters straight up from the lunar surface and standing next to what appears to be a rectangular shadowed hole or depression running from its base out to the right.

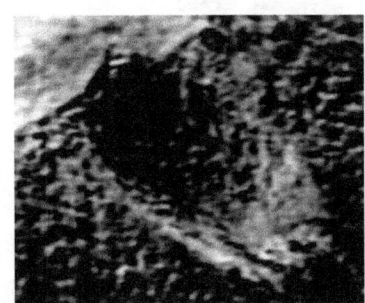

Apollo-16 AS16-P-5024 Enlarged

207

Lobachevski impact craterlet enlarged. Shows sun reflecting off
bottom and right side of craterlet rim. There is no "tower" object.
What is visible are boulders and ejecta debris.

This is impossible. The original photograph (AS16-121-19407)
shows the lunar surface sloping down to the left and the
enhancements are likewise orientated. If the lunar horizon is
properly orientated to an observer standing on the surface, it
becomes evident that the illusive 'spire' is actually leaning at a
considerable angle, possibly as much as 45°, to the observer's right.
It is possible that some structure can exist at such an extreme angle,
but the probability is almost zero in light of all the facts presented
here.

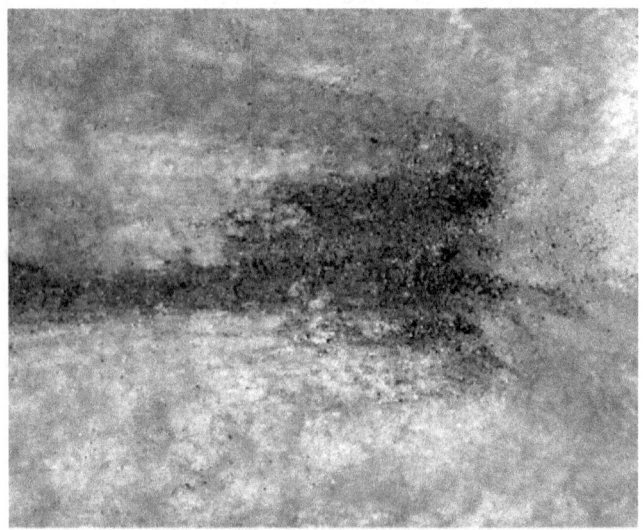

LROC Photo M176899195LC
Crater Lobachevski from a direct 'top view'. No object seen.

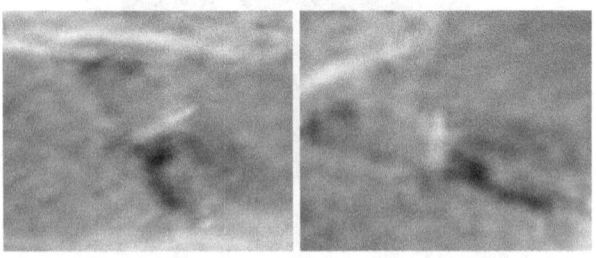

Craterlet
LEFT: Enlargement as done by graphics expert.
RIGHT: Enlargement of true angle – shows highlight at 45%

As for photographic enhancements, pixel streaking is visible nearby, aligned with the vertical axis of the picture, and parallel to the axis of the 'spire'. This is particularly noticeable and suggests that the 'spire' is an artifact of excessive imaging and enhancement processes.

There is no excuse for persons such as Mss X. Everyone has access to computers, the Internet, imaging software's and NASA photos, and with very little effort and download time good quality pictures of these type sites can be obtained.

As for Mr Leonard's analysis, we might allow him some grace in his speculations, for lack of computer technology and recent higher resolution photos. Nevertheless, he did have access to higher resolution photos of Crater Lobachevski and development laboratories! Pray, maybe we should put the blame on his lack of funding rather than on his lack of credibility. Not really. A good positive print from a sharp negative was more than affordable in those days just as it is today.

30
EXAMPLES OF CRATER RIM LANDSLIDES

The above Lobachevski crater rim collapse is most probably a result of a large micrometeoroid impact. Crater Kant P is an example of crater disturbance rim-slide; caused by a number of reasons, primarily large micrometeoroid impact, where part of the rim collapses downward toward the center of the crater. Kant P lies in the central highlands on the Moon's far side.

The younger, small pear-shaped crater on Kant P's north wall is an excellent example of the controlling effect that topographic relief plays on the shape of an impact crater. Because the small crater was formed on a steeply sloping surface, the ejecta deposited chiefly down-slope and formed as broad rim. The original rim and wall on the up-slope side is obliterated by slumping. The slumping has left a landslide scar and has caused the deposition of talus and scree in the lower part of the crater [24].

At the right angle, the same sun highlight anomaly would develop and present an artificial "tower" looking object. The small impact crater rim of brighter ejecta material would similarly give off a super bright highlight and give the appearance that there was an object sitting there.

When the La Pérouse-A impact excavated material from the side of a hill of highland material (1.5 to 2 km taller than the plains to the west), the slope became unstable and collapsed into the crater La Pérouse A. The result is a crater with one rim 920 meters higher. The Featured Image shows the top of this landslide, now the new rim of La Pérouse A. The landslide material inside the rim is high reflectance, while the undisturbed section of the highland material is relatively darker.

Despite being the same composition, the landslide material is higher reflectance since it is fresh (recently uncovered). Faint lines along the outer edge of the rim are evidence of highland material slumping towards the crater. Collapse features on the Moon often display slump lines, which show where material has fractured and moved downwards, but not completely collapsed. Subsequent impacts or seismic shaking could cause further landslides inside the crater! **[25].**

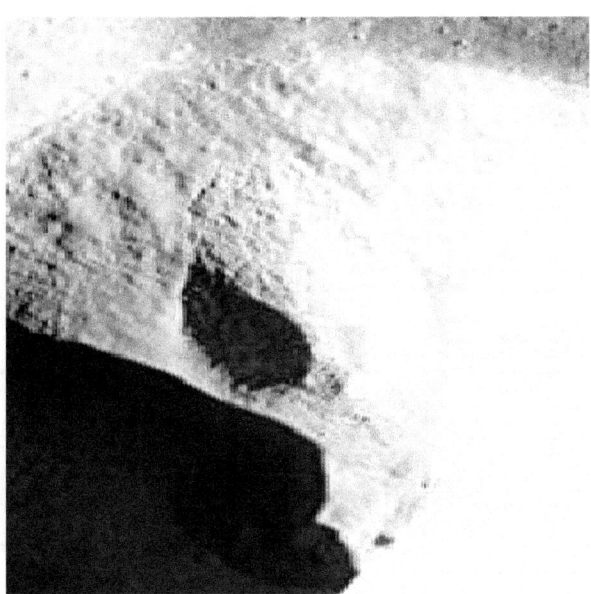

AS16-P-4559_FULL_MED
Crater Kant -P impact crater landslide, rim collapse slide. La Pérouse A Crater IMPACT LANDSLIDE

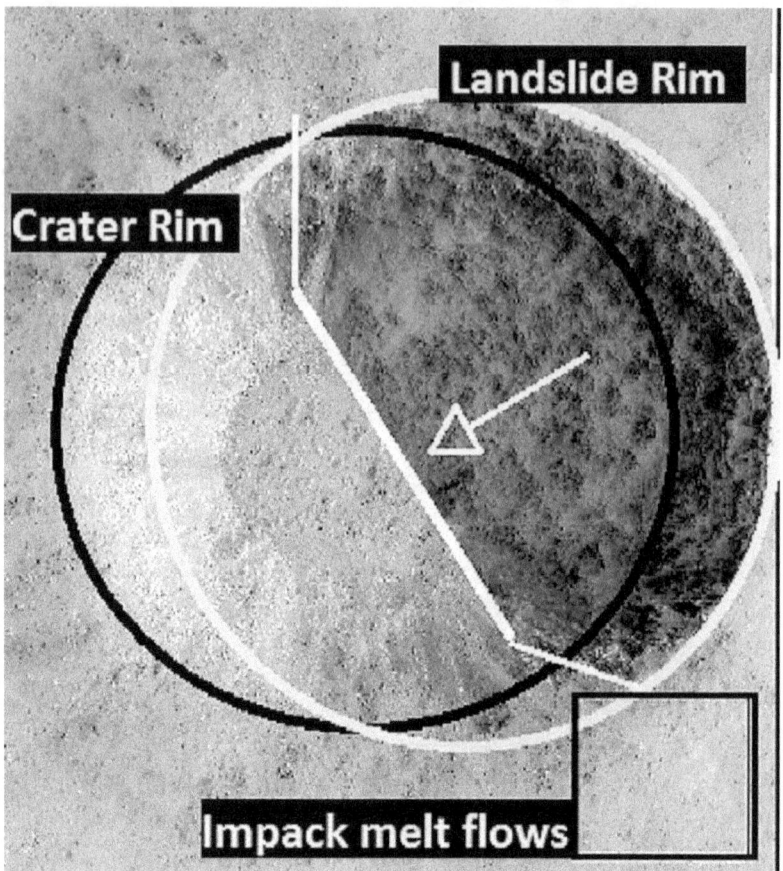

The yellow (white) line shows the crater rim and landslide, while the black dashed line shows the new rim due to the landslide. Reduced scale NAC mosaic (M152390311LR), image is ~4.5 km across [NASA/GSFC/Arizona State University].

TRYING TO BRIDGE THE GAP!
CREDIBILITY IS MARE'D

A supposed close friend and supporter of George Leonard and the aliens on the moon theory is Vito Saccheri. Like George, who inspired his alien presence delusion, he also claims to have seen classified lunar photographs at NASA in the 1970's showing all sorts of ET constructions.

He said he observed *"A boulder that seemed to have been rolled uphill, leaving its tracks in the side of the hill; obvious machinery on the surface, showing bolted sections; three dilapidated "bridges" crossing a chasm that reminded me of the Grand Canyon; pipe fittings that looked like four-way T's (or X's) that could be seen in every photo, some with their ends turned up or down as they hung over the edge of a crater; three surprising pyramids... apparent pipelines cris-crossing the surface, running to and from craters; a UFO rising from the surface and photographed directly above a crater; and perhaps the most memorable, the unmistakable figure of a rectangular structure placed squarely in the biggest crater pictured- the structure looked either very old or under construction, but the crater had to be miles wide, and the camera angle gave a perfect three-dimensional view"* [26].

Unlike the Lunar Orbiter Cameras, let us zoom in on the credibility of Vito's claims. Apparently, his claim as to what he saw in the NASA photo collection is no more credible than his story on how he came to see them.

Saccheri recounts a 1979 experience with a Lester Howes in Houston. After they met, he said, Mr Howes, an amateur astronomer and a ufologist, wanted help in obtaining access to secret (classified?) photos held somewhere inside NASA. Saccheri claims: *"Les showed me a small paperback book entitled "Somebody Else Is on the Moon," written by a former NASA scientist, George H. Leonard. Leonard had been working in the photo intelligence division of NASA."*

Is this pure journalism or what? The information is not true at all, for Leonard never worked for NASA, especially the 'photo Intel' division. Leonard never CLAIMED he worked for NASA.

Mr Saccheri claims that after Leonard's' arguing futilely with NASA authorities about releasing the photos, only then was he able to publish them in his book. Again, this is a lie, for Leonard GOT the photographs from published sources – sources everyone has access to. He did NOT ever possess "unpublished" photos.

The Houston Sky continues to point out: *"Reinforcements were called in, and we soon found ourselves having the same conversation with the big boys from administration. None had seen the book, but significantly, one had taken the time to confirm that Leonard was in fact a former NASA scientist-at the Jet Propulsion Lab, from what I could gather."*

Good God! Besides no NASA employee of any quality and integrity would want to be seen with such a comic book, Saccheri blows the fictional account cover when he claims George worked for NASA. But, George Leonard never worked ANYWHERE for NASA and never claimed that he worked for NASA.

As to the mysterious hidden, hard to get to, offbeat, out of the way NASA photo archive, Saccheri continues to lie: *"We were directed to a Building 30, which had not been on the tour and which didn't even exist. Building 30-A turned out to be empty so we walked into Building 30-B and found ourselves in the middle of a high-security area where an existing mission was being monitored. Finally, some serious discussions transpired. The photo library, we were told, had been relocated off site to the 'Lunar Landing Observatory' directly adjacent to the east NASA property on NASA Road 1."*

This is also pure space spam. Building 30 existed, and was on the public tours (it is better known as 'Mission Control'), and was easy to visit, because in 1979 there WERE no NASA manned space missions to monitor. The lunar photographs were (and always have been) in the "Lunar Science Institute", the former Jim ("Silver Dollar") West hunting lodge, east of JSC on NASA Road 1. The name and the 'relocation' are erroneous.

Saccheri continues to fictionalize that *"Two days later, we drove east on NASA 1 past the main entrance of the facility, found a chain-link fence that marked the eastern limit of the property line, expecting to see a building or sign. Nothing but a heavily wooded area! Driving back and forth along the road trying to decide whether they had done it to us again, we noticed a narrow dirt road running back into the woods directly along NASA's fence line. Hung on the chain between two small posts was a sign that read simply: 'No Trespassing.' Instinct told us this had to be the place. We lowered the chain and drove about three-quarters of a mile down the dirt road, which U-turned back toward the highway. Directly behind the trees and camouflaged by the woods was our building.*"

This is ever so much bull and chicken droppings, for the building is 50 yards from the road, with an open lawn in front of it. The driveway is paved. Anyone can easily visit the place. There was no chain across any dirt road – not until years later when the institute moved to a new, much larger building about two miles to the north.

And this fairy tale goes on with, "*When we told the receptionist we wanted the library, she pointed toward the broom closet, which as it turned out, opened onto a winding stairway leading down into a dimly lit under-ground tunnel. I'm certain it took us back toward the NASA property line.*" Sounds like a scene from some Indiana Jones adventure. The truth is the LSI had no basement, because it was all crawl space. The photo archives were in the library just off the large entry hall – on the ground level.

When it comes to technical details, the lie gets worse: "*He [the fictional librarian at NASA] explained that for security reasons, NASA had split the country into five regions, each with a duplicate set of records and a different code number system. Leonard's numbers weren't applicable in this facility. I asked where the master list was kept, and Roger [the fictional librarian] replied, 'at Langley, Virginia.' Les and I looked at each other.*"

This final dose of sleeping potion is the ultimate in misinformation: The NASA Apollo photographs have exactly one code number, but when reissued as part of press releases, they received a public affairs office reference number too. These were the numbers Leonard used, which shows that the pictures had been published. The press office has a conversion sheet for the numbers it uses.

Of course, the Houston Sky article also mentions that NASA would not let him take notes, pictures or make copies of the

negatives – a convenient way of Saccheri hiding the fictional BS!
[27]

NSSDC Lunar Photo Archives

32
CONCLUSIONS:

Is there really someone else on the moon? Did we find artificial lunar (boulder) crop circles or over blown boulder balderdash? Were there tools in Tranquility or was it all tranquilizing tall tales of Tom Foolery? Were there found any alien crater condos for sale or should we crater the alien condo con job?

Lunar highways and byways or by the way, imagine these highways? We found no technology in Tycho, but only Leonard's fantasies and journalist typos. The alien made rays are not real. They are only George's personal speel. There are no wheels, no axles, no metal or steal, no cuts or crevasses, no ditches or stitches; no eggs, ovals, hedges or wedges. No nuts or bolts and no hammers or screws; no crescents, no crosses, mine pits or food fields. We find not water, not a shower, not cuspids or tower, no Egyptians or Mayans, no pyramids or any geometric angle, no artificial things that mysteriously dangle; no cars and no tires, no gears or spires. We cannot find rings, domes or any giant speakers, let alone skulls, mice, ducks or people. Bridges and causeways really do not exist, no more than junkyards. X-drones and dirt rovers are but pixel glitches, while there are no dinosaurs, eagles, dots, dashes, and super-size rigs and hitches.

Sorry, there are no giant treads, tractor trails or ladders, no numbers or letters, no missiles or rocket towers. When people try to make mountains out of a molehill and tell lies to bridge the gap, time and reason soon expose it as just pure flap. No, aliens have not visited our solar system.

FOOTNOTES

SECTION-I
INTRODUCTION

(1a.) See, "NASA's Alien Anomalies caught on film - A compilation of stunning UFO footage" (http://www.youtube.com/watch?v=WlLN_Jcg1pc).

[1]. Letter to the Ecclesia in Rome, Chapter 11, verse 36 and Col. 1:16-17).

[2]. The Philosophical Impossibility of Darwinian Naturalistic Evolution by
Dennis Bonnette, Ph.D.
www.godandscience.org/evolution/philosophy_darwinian_evol
ution.html].

[3]. "The Exoplanet Next Door: Nature News & Comments". Nature.com. doi:10.1038/nature11572

[4]. Wikipedia: "Rocket Equation" and "Russians to ride a nuclear-powered spacecraft to Mars: See,
www.csmonitor.com/World/Global-News/2009/1029/russians-to-ride-a-nuclear-powered-spacecraft-to-mars

[5]. Wikipedia: "Nuclear thermal rocket"

[6]. D.F. Spencer and L.D. Jaffe. "Feasibility of Interstellar Travel." Astronautica Acta. Vol. IX, 1963, pp. 49–58.

[7]. General Dynamics Corp. (Jan. 1964). "Nuclear Pulse Vehicle Study Condensed Summary Report."

[8]. Ross, F.W. – Propulsive System Specific Impulse. General Atomics GAMD-1293 8 Feb. 1960
And, Eugen Leitl: Advanced nanotechnology - 3 new articles. Postbiota.org, December 1, 2007, retrieved July 18, 2011

(8a) Wikipedia, under "Project Orion."

[9]. "Destructive Physical Analysis of Hollow Cathodes from the Deep Space 1 Flight Spare Ion Engine 30,000 Hr. Life Test". Retrieved 2007-11-21. And, "The Prius of Space", September 13, 2007, NASA Jet Propulsion Laboratory.

[10]. "Dyon-fermion dynamics" Curtis G. Callan, Jr. 1982, and "Searches for Proton Decay and Superheavy Magnetic Monopoles" B. V. Sreekantan, 1984.

[11]. Wikipedia: "Bussard ramjet"

[12]. Wikipedia: "Beam Propulsion" and "Robert L. Forward"

[13]. See, "God and ETs", http://www.answersingenesis.org).

[14]. Wikipedia: "Interstellar Travel"

[15]. O'Neill, Ian (Aug. 19, 2008). "Interstellar travel may remain in science fiction". Universe Today.

[16]. "RADIATION HAZARD OF RELATIVISTIC INTERSTELLAR FLIGHT" By Oleg G. Semyonov. p. 2

[17]. "Starship pilots: speed kills, especially warp speed." Concepts by Valerie Jamieson, February 2010. www.newscientist.com

[18]. Wikipedia: "Telomeres"

[19]. Hazards of Spaceflight. Hubpages.com

[20]. RADIATION CANCER RISKS AND UNCERTAINTIES FOR DIFFERENT
MISSION TIME PERIODS. Myung-Hee Y. Kim and Francis A. Cucinotta. COSPAR, 2012

[21]. Semyonov, p. 4

[22]. Semyonov, p. 5

[23]. Semyonov, p. 13-14

[24]. Wikipedia, under "Moon rock"

[25]. James Papike, Grahm Ryder, and Charles Shearer (1998). "Lunar Samples". Reviews in Mineralogy and Geochemistry 36: 5.1–5.234

[26]. Wikipedia, under "Lunar soil"

[27]. Ibid. paragraph 8

[28]. Taylor, Gerald. p. 1939. "Microbial assay of lunar samples" Proceedings of the Second Lunar Conference, Volume 2, pp. 1939-1948.; The M.I.T. Press, 1971

[29]. "The Sagan Criteria for Life Revisited." http://spacescience.spaceref.com/newhome/headlines/ast21may99_1.htm

[30]. Heiken et al. Lunar Sourcebook: A User's Guide to the Moon. Cambridge: University of Cambridge Press, 1991. See Section 8.8.

[31]. THE LUNAR SOURCEBOOK, p.634

[32]. Luna 16 and Luna 20 Drill, by David R. Williams, NASA Goddard Space Flight Center

[33]. Micro-organisms from the Moon discovered by the Russians in 1970

[34]. Many times, evolutionists have triumphantly announced 'proofs' of evolution against the Christian worldview, and the secular media uncritically broadcast them over the airwaves.

However, later, further discovery discredits the evidence. We have only to remember Archaeoraptor, pushed as 'proof' of dinosaur-to-bird evolution by the influential National Geographic, but later exposed as a hoax (c). More closely related to the moon monuments claim is of course the alleged 'fossilized life' found in the meteorite ALH84001, supposedly from Mars. These claims are retracted later for a number of reasons, yet the media does not give retractions much publicity

See, (d) "Life on Mars?" Separating fact from fiction and Mars claims weaken further.

[35]. Astrobiology: Biology Cabinet. September 26, 2006. 2011-01-17

(36). (A) Wikipedia, under "Life on Mars"

(37). Wikipedia, under "Mars soil" and "Phoenix lander"

(38). http://www.nasa.gov/mission_pages/msl/news/msl20120927.html

(39). http://www.cnn.com/2006/TECH/science/06/02/red.rain/index.html . The research papers of Dr. Godfrey Louis are available in the Los Alamos National Laboratory (LANL) archives.

(40). (A) Colored Rain: A Report on the Phenomenon. By S. Sampath, T.K. Abraham, V. Sasi Kumar and C.N. Mohanan, Centre for Earth Science Studies, Thiruvananthapuram 695031, India, Tropical Botanic Garden and Research Institute, Pacha, Palode, Thiruvananthapuram, India, November 2001.

(41). Courtesy: BBC News Online. http://news.bbc.co.uk/2/hi/science/nature/1466477.stm

(42). Precambrian Research 106 (2001) 15–34. "Life on Mars: evaluation of the evidence within Martian meteorites ALH84001, Nakhla, and Shergotty" by E.K. Gibson, Jr., and associates.

(43). Meteoritics and Planetary Science, 2007. Dr. Bell continues her meteorite research at NASA, where she has worked since 1994, curating meteorite samples collected by NASA teams in Antarctica. See, http://breakthrough.nsm.uh.edu/2008_11/bell.htm

(44). New Scientist 165(2228):21, March 4, 2000

(45). "Mars Meteorite Not Evidence of ETs" by Charles Q. Choi. See, http://www.livescience.com

(46). Once upon a time with Dr. Rhawn Joseph, in "Life and Death on the Red Planet". http://brainmind.com/Mars.html

(47). Dr. Rhawn Joseph, Journal of Cosmology. http://journalofcosmology.com/Life101.html#1

(48). SEDIMENTARY HEMATITE ON MARS AND ITS IMPLICATIONS FOR THE EARLY MARTIAN ENVIRONMENT. D. C. Catling and J. M. Moore. Lunar and Planetary Science XXXII (2001)

(49). (A) A formation mechanism for hematite-rich spherules on Mars. By Chaojun Fana, Hongjie Xiea, Dirk Schulze-Makuchb and Steve Ackleya. Department of Geological Sciences, the University of Texas at San Antonio, TX.

(50). (A) Dr. Rhawn Joseph, in "Life and Death on the Red Planet" and from, http://brainmind.com/Mars.html

(51). "Abiotic synthesis of polycyclic aromatic hydrocarbons on Mars" by Mikhail ZolotovEverett Shock. Journal of Geophysical Research: Planets. (1991–2012). Volume 104, Issue E6, pages 14033–14049, January 1999. FULL QUOTE:

"Thermochemical calculations of metastable equilibria are used to evaluate the stability of condensed polycyclic aromatic hydrocarbons (PAHs) in cooling thermal gases and hydrothermal fluids on ancient Mars, which are roughly similar to their terrestrial counterparts. The effects of temperature, pressure, the extent of PAH alkylation, and the relative stability of PAHs and alkanes are considered. Inhibition of methane and graphite formation favors synthesis of metastable mixtures of hydrocarbons from aqueous or gaseous CO, CO_2, and H_2 below 200°–300°C. High-temperature quenching of H_2 and CO in volcanic and impact gases and dynamic hydrothermal fluids also favor the synthesis of hydrocarbons. In addition, an excess of CO in cooling systems relative to equilibrium makes the synthesis from CO and H_2 more favorable energetically than from CO_2 and H_2. Both the CO-H_2 reactions through Fischer-Tropsch (FT) type processes and the CO_2-H_2 reactions could be catalyzed by magnetite. Volcanic gases and hydrothermal fluids related to mafic and ultramafic magmas and rocks are more favorable for FT type synthesis than those associated with oxidized Fe_2O_3-bearing rocks and regolith. We conclude that PAHs and aliphatic hydrocarbons on Mars and

Earth could be formed without the contribution of biogenic carbon. Some PAHs could be formed because of pyrolysis of other hydrocarbons formed earlier by the FT type synthesis or other processes. If the PAHs found in the ALH 84001 Martian meteorite formed together with other hydrocarbons through FT type synthesis, it may be possible to bracket the temperature of the synthesis."

(52). "No, PAHs do not mean life on Mars – again and again." Science Magazine. 2012/05/23

(53). NASA Scientist Discovers Alien Life in Meteorites - Again! NOT! By Rich Deem. http://www.godandscience.org

(54). Gary Bates: http://creation.com/did-god-create-life-on-other-planets

(55). Allan H. Treiman, Lunar and Planetary Institute, Houston, Texas. FULL QUOTE: *"Purported biogenic features of the ALH84001 Martian meteorite (the carbonate globules, their submicron magnetite grains, and organic matter) have reasonable inorganic origins, and a comprehensive hypothesis is offered here. The carbonate globules were deposited from hydrothermal water, without biological mediation. Thereafter, ALH84001 was affected by an impact shock event, which raised its temperature nearly instantaneously to 500-700K, and induced iron-rich carbonate in the globules to decompose to magnetite and other minerals. The rapidity of the temperature increase caused magnetite grains to nucleate in abundance; hence individual crystals were very small. Nucleation and growth of magnetite crystals were fastest along edges and faces of the precursor carbonate grains, forcing the magnetite grains to be platy or elongated, including the "truncated hexa-octahedra" shape. ALH84001 had formed at some depth within Mars where the lithostatic pressure was significantly above that of Mars' surface. Also, because the rock was at depth, the impact heat dissipated slowly. During this interval, magnetite crystals approached chemical equilibria with surrounding minerals and gas. Their composition, nearly pure Fe_3O_4, reflects those of equilibria; elements that substitute into magnetite are either absent from iron-rich carbonate (e.g., Ti, Al, Cr), or partitioned into other minerals during magnetite formation (Mg, Mn). Many microstructural imperfections in the magnetite grains would have*

222

annealed out as the rock cooled. In this post-shock thermal regime, carbon-bearing gas from the decomposition of iron carbonates reacted with water in the rock (or from its surroundings) to produce organic matter via Fischer-Tropschlike reactions. Formation of such organic compounds like polycyclic aromatic hydrocarbons would have been catalyzed by the magnetite (formation of graphite, the thermo chemically stable phase, would be kinetically hindered)."

(56). NASA: "The Conditions for the Emergence of Life were Present on Mars — Period, End of Story" (Today's Most Popular) February 17, 2012. See, http://www.dailygalaxy.com/my_weblog/2012/02/nasa-the-conditions-for-the-emergence-of-life-were-present-on-mars-period-end-of-story-todays-most-p.html

(57). "The Enigma of Methane on Mars." http://exploration.esa.int/science-e/www/object/index.cfm?fobjectid=46038f

(58). "Curiosity's detection of methane on Mars could suggest ET life." Jason McClellan, http://www.openminds.tv/curiositys-detection-of-methane-on-mars-could-suggest-ET-life-865

(59). "No Methane on Mars." Shawn Domagal-Goldman, http://www.astrobio.net/paleblueblog/?p=1689

(60). "Mars Methane? NASA's Curiosity Rover Finds None--Yet." Mike Wall: 11/02/2012 02:42 PM, SPACE.com

(61). Condensed from "Biology Cabinet Organization" by Biol. Nasif Nahle., http://www.biocab.org/Mars.html

(62). Ibid. http://www.biocab.org/LifeOnMars.html

(63). "Did Life Come from Outer space? http://creation.com/did-life-come-from-outer-space.

(64). Nature 404(6779):700, April 13, 2000. And, Boyd, R., "Sorry, but we are alone", The Courier-Mail, Brisbane, Australia, April 14, 2000, p. 10

(65). Hawking, Steven. "The Grand Design" and Dr. John Lennox, notes 11, 12, 13.

(66). Ref. http://www.changinglivesonline.org/evolution.html

(67). Newman, 1967, p. 662

(68). http://www.christiananswers.net/q-crs/abiogenesis.html

(69). Five Arguments against the ET Origin of Unidentified Flying Objects JACQUES F. VALLEE. Journal of Scientific Exploration, Vol. 4, No. 1, pp. 105- 1 17, 1990

(70). Wikipedia: Under "Incubus"

(71). "On the likelihood of non-terrestrial artifacts in the Solar System"

By Jacob Hagg-Misra and Ravi K. Kopparapu, Nov. 2011

(72). Ref. "Scientists Suggest Moon Photos May Reveal ET Visitation."

http://www.pakalertpress.com/2012/01/04/scientists-suggest-moon-photos-may-reveal-ET-visitation

SECTION-2
GEORGE LEONARD
FOOTNOTES:

[1a] http://en.wikipedia.org/wiki/Olivineolivine,
http://en.wikipedia.org/wiki/Pyroxenepyroxene,
http://en.wikipedia.org/wiki/Plagioclaseplagioclase,
http://en.wikipedia.org/wiki/Ironferro-
http://en.wikipedia.org/wiki/Titaniumtitanic
http://en.wikipedia.org/wiki/Oxideoxide
http://en.wikipedia.org/wiki/Ilmeniteilmenite.

(1b) Rukl, Antonin, Atlas of the Moon.

(1c) A New Photographic Atlas of the Moon, by Zdenek Kopal and Harold C. Urey (1971), Plate 211, page 290.

[1]. The Lunar Orbiter program photographed 99% of the surface of the Moon with resolution down to 1 meter. The Lunar Reconnaissance Orbiter Camera (LROC) consists of one Wide Angle Camera and two Narrow Angle Cameras to provide high-resolution images (0.5 to 2.0 m pixel scale) of key targets - a pixel scale close to 75 meters (246 feet). A single measure of elevation (one pixel) is about the size of two football fields placed side-by-side.

[2]. The origin of lunar crater rays: B. Ray Hawke. www2.ess.ucla.edu].

[3]. Location of Blair Cuspids in LROC. Center latitude 5.07, Center longitude 15.57. And, "Lunar Rocks," Ch. 6 - G. Jeffrey Taylor.

(3a). H.L. Hayward, Symbolic Masonry: An Interpretation of the Three Degrees, Washington, D.C., Masonic Service Association of the United States, 1923, p. 207; 'Two Pillars' Short Talk Bulletin, Sept., 1935, Vol. 13, No 9; Charles Clyde Hunt, Some Thoughts On Masonic Symbolism, Macoy Publishing and Masonic Supply Company, 1930, p. 101

(3b). W. Wynn Wescott, Numbers: Their Occult Power and Mystic Virtues, Theosophical Publishing Society, 1902, p. 33

(3c). Haywood, quoted above, p. 206-7 and Rollin C. Blackmer, The Lodge and the Craft: A Practical Explanation of the Work of Freemasonry, St. Louis, The Standard Masonic Publishing Co., 1923, p. 94

[4]. Washington Post 11/23/66 and the Houston Sky, No.5, June/July 1995

[5]. Mysterious "Monuments" on the Moon. Argosy Magazine, August, 1970 ABAKA = a triangle. NOTE: Nowhere have I found the word "abaka" or "abaca" except when used for a Tropical plant native to the Philippines grown for its textile and papermaking fibre (Musa textilis) also called Manila hemp. Maybe Mr Blair has smoked too much and sees too much in these piles of rocks?

[6]. The Boeing News article, March 30, 1967

[7]. LO-III-214-M, a medium shot of Planitia Descendus the hopeful landing site of Luna 9.http://www.moonviews.com/archives/2009/06/lunar_orbiter_ima ge_recovery_p_5.htmlwww.moonviews.com/archives/2009/06/lun ar_orbiter_image_recovery_p_5.html

[8]. "New LO-IRP high res Lunar Orbiter image of western Oceanus Procellarum." Post Wed. June 10, 2009, Joel Raupe. Lunar networks.

[9]. "Boy, that sure looks like Luna 9!" Post Dec. 3, 2011, Joel Raupe. Lunar networks.

[10]. Moon Morphology, Shultz, p. 14

[11]. Conceptual Design of a Fleet of Autonomous Regolith Throwing Devices for Radiation Shielding of Lunar Habitats. Dennis Wells, 1992. NASA-CR-192078. P.10-11

[12]. Dome cities for extreme environments." Leonard, Raymond S.; Schwartz, Milton. Third SEI Technical Interchange: Proceedings.

[13]. a.) "The Sminthian Apollo and the Epidemic among the Acheans at Troy," by Frederick Bernheim and Ann Zener. b.) "The Cults of the Greek states. Vol.4. by Lewis Fernell. c.) "Pestilence and Mice," by James Moulton and A.T.C. Cree.

d.) "Apollo Smintheus, Rats, Mice, and Plague," by A Lang.

[14]. Apollo Moon Conversations. www.ufos-aliens.co.uk).

[15]. Wilkins, 1954

[16]. Evidence for Artificial Structures on the Moon. Lan Fleming, 1995

[17]. "GEOLOGIC MAPPING OF THE KING CRATER REGION WITH AN EMPHASIS ON MELT POND ANATOMY - EVIDENCE FOR SUBSURFACE DRAINAGE ON THE MOON." J. W. Ashley.

[18]. "LROC finds natural bridges on Moon." See, http://lunarscience.nasa.gov/articles/lroc-finds-natural-bridges-on-moon/http://lunarscience.nasa.gov/articles/lroc-finds-natural-bridges-on-moon/ and, http://lroc.sese.asu.edu/news/index.php?/archives/277-Natural-Bridge-on-the-Moon!.htmlhttp://lroc.sese.asu.edu/news/index.php?/archives/277-Natural-Bridge-on-the-Moon!.html#extended

[19]. "The Mare Tranquillitatis PIT" http://lunarscience.nasa.gov/articles/lroc-captures-high-sun-view-of-mare-tranquillitatis-pit-crater/http://lunarscience.nasa.gov/articles/lroc-captures-high-sun-view-of-mare-tranquillitatis-pit-crater/

[20]. "The PIT" http://www.nasa.gov/mission_pages/LRO/news/first-year.htmlhttp://www.nasa.gov/mission_pages/LRO/news/first-year.html

[21]. NASA SP-168 "Exploring Space with a Camera." John McCauley of USGS, p.86

[22]. Wikipedia: "Ray system"

[23]. Hubble photograph of Moon pyramid.
http://alienanomalies.activeboard.com/f494751/lunar-and-iss-anomalies/
Also see
http://www.godlikeproductions.com/forum1/message2097592/pg1]
.

[24]. Apollo over the Moon: p.__

[25]. LROC. Frame M152390311LR
http://lroc.sese.asu.edu/news/index.php?/archives/633-Top-of-the-Landslide.html

[26]. NEXUS Magazine. Oct-Nov. 1995. p.47.

[27]. When Mr Saccheri mentioned that the librarian said that the lenses of the Lunar Orbiter cameras could zoom in on an anomaly when the on-board computer ran across a funny looking surface feature and snap high resolution frames, he was either outright lying or the (fictitious) librarian was ignorant. The on board computers had tiny RAM, and the lenses were unable to zoom in. This whole story is obviously a historical fiction promoted by an alien presence profiteer.

PHOTOS:

The following listed PHENOMENA descriptions were taken from the book "Somebody Else is on the Moon". NASA plate numbers and plate descriptions are freely available and listed on all NASA public moon plate photographs.

APPENDIX - 1
Leonard's NASA photo numbers, phenomena, and lunar locations - Cross-referenced to Johnson Space Center LO Numbers:

66-H-1293 lunar far side octagonal crater L. Orbiter I
= LO-I-1136-H3

66-H-1611 Western Mare Tranquillitatis

66-H-1612 South-eastern Mare Tranquillitatis
= LO-II-42-H1-3?

67-H-41 (plate-27) Mare southeast of crater Kepler
= LO-II-182-H1

67-H-187 (plate-28) Lunar Orbiter III
= LO-III-012-H1

67-H-201 Crater Kepler in Oceanus crater
= LO-III-162-M and "H1"

67-H-266 anomalous crater Surveyor I landing site

67-H-304 South of Maskelyne F

67-H-318 Oceanus Procellarum

67-H-307 West of Rima Maskelyne in Southern Mare
Tranquillitatis

67-H-327 crater in Oceanus Procellarum

67-H-510 Crater Sabine D

67-H-758 Cratered upland basin taken Lunar Orbiter II
= LO-II-62-H3 and "M"

67-H-897 Northeast of Mare Imbrium near Alpine Valley
= LO-IV-115-H3

67-H-935 Mare Orientale. Mare Veris and Rook Mountains
= LO-IV-187-H2

67-H-1135 Crater Vitello with Boulder Track
= LO-V-168-H2

67-H-1179 Tycho crater Tycho

67-H-1206 Tycho Crater

67-H-1400

67-H-1409

67-H-1651 Crater Tycho and northern highlands

67-H-8 lunar far side taken from Apollo 8
= AS8-13-2244

69-H-25 unnamed far side crater

69-H-28 Crater Humboldt and Southern Sea surrounding craters

69-H-737 Triesnecker crater

69-H-1206 Tycho Floor
= LO-IV-125-M

70-H-1629 Fra Mauro area

70-H-1630 Fra Mauro area

70-H-781 Taken by Apollo 14

70-H-1300

70-H-1765 Oceanus Procellarum and Herodotus

70-H-834 North-western half of King Crater
= AS16-120-19263--19267

72-H-835 Mare Crisium, Mare Tranquillitatis and Crater Proculus

72-H-836 Far side King crater highlands
= AS16-120-19226

72-H-837 King Crater

72-H-839 Far side King Crater
=AS16-120-19228

72-H-1109 East of Mare Smythii

72-H-1113 Northwest of King Crater

72-H-1387 Lubinicky area
= LO-IV-125-H2

PHOTOS AND ILLUSTRATIONS:

Picture of George Leonard's Book: "Somebody Else Is on the Moon" Mare Tranquillitatis. Latin for Sea of Tranquillity: Named in 1651 by the astronomers Francesco Grimaldi and Giovanni Battista Riccioli in their lunar map 'Almagestum novum'
Lunar Orbiter II-042-H1 "The Rock Pile" Rukl's "Moon Viewed by Lunar Orbiter (p.49)9 NASA SP-206 Lunar Orbiter Photographic Atlas of the Moon.
[www.lpi.usra.edu/resources/lunar_orbiter/book/lopam.pdf]
LO-II-042-H1 George's sketch of 66-H-1612 "Oval" shaped vehicles (p. 101)
(Kopal, Pl. 211, page 290)
NASA 67-H-304 A sharply defined, machine-tooled object is indicated in the area of Crater Maskelyne-F. (p. 177)
NASA PHOTOS:

PLATE 26, 70-H-1629, PLATE 26, Diggers, Dozers and Dirt-flingers.

LO-II-182-H1, 67-H-041 (Plate 27)

-H-1630 Plate 24, Control Wheels.

-H-1611 More crater wheels.

Surveyor 7. Crater Tycho Photomosaic. The central horizon hills are eight miles away from the spacecraft.

V-125-M Tycho.

LO-III-194-H2 More craters with devices and debris. Possible "Play-boy Bunny Rabbit" face in bottom right picture.

Crater Tycho LO-V-125-M

Crater Tycho Photomosaic taken by Surveyor 7. The central horizon hills are eight miles

away from the spacecraft.

Tycho crater's central peak complex, shown here, is about 9.3 miles (15 km) wide, left to right (southeast to northwest in this view). (LROC); and TYCHO CRATER CENTRAL PEAK View of the summit area of Tycho crater's central peak. Boulder in the background is 400 feet (120 m) wide. The image itself is about 3/4ths of a mile wide. (Credit: NASA Goddard/Arizona State University)

69-H-1206 Leonard's Sketch, p.61 Some munched on pie chunks left by aliens.

LROC M190672344RC_pyr Lava flows and ground disturbances with collapsed crater rim.

Large picture of Crater Tycho and crater rays, medium shot.

Octagonial "oval" shaped covering with "PAF" glyph lettering. (p.120)

Leonard's sketches of "oval" spacecraft's. LO V-125-M and H2, and the LROC, Frame M129369888LC_pyr

LO V-125-H2 compared with LROC. Supposed "screw" in white box is not there.

Crater Tycho.

NASA 67-H-1651. An object in the Highlands, north of Tycho. (p. 124)

Tycho Floor A and B

LO-V-127-M. The only big "screw" observable in crater Tycho.

LO-V-126-M

LO-V-125-H2

LO-V-126-H2

Fingals_Cave_Staffa_Scotland. Organ-pipe type basalt lava.

Sleeping Pele is an example of "twisted lava" flow on Big Island, Hawaii.

Tycho Floor disruptions: Crater Tycho is littered with "alien" looking geometric shapes - all formed out of the geothermal disfiguring of the crater floor.

The BIG SCREW Photo and Leonard's sketch of Crater Tycho "screw device", p. 124

Ibid

Fingals_Cave_Staffa_Scotland. Organ-pipe type basalt lava.

Sleeping Pele is an example of "twisted" lava flow on Big Island, Hawaii; and More LAVA Shapes in Earth Nature.

LO-III-162-M (NASA 67-H-201) An oblique view of crater Kepler.

THE CRATER KEPLER (LROC).

[nasa.gov/images/content/513108main_012511b.jpg]

Craters with crosses and regolith sprayings, p. 132. NASA 67-H-201

LO3-162-M Leonard's Latin shaped cross lying at angle. (p.62) Crater Kepler area objects outlined.

BLAIRE CUSPIDS: Photo of area of RIMA ARIADAEUS and Crater Silberschlag.

Location topographic map for LO-II-62-M

LO-II-62-M

Enlargement of LO-II-62-M, NASA 86-H-758; LO-II-62-H3 (Negative image) Enlargements of medium shot.

LROC M159847595RC_pyr, Two small craters. BLAIR CUSPIDS (Closeup-1)

And, LO-II-062-H3 Cuspids in Aeriadaeus B, and large shadow on rim close-up.

-48 LUNA-9 photo of tower and shadow.

LO-III-67-H-187 and Leonard's sketch, p. 169.

LO-IRP Restored version of LO-III-214-M. Restored image of Planitia Descendus

LO-III-214-M

M132071202LC_pyr. Planitia Desscentus "Plain of Descent," borderland along the western edge of Oceanus Procellarum. Supposed Luna-9 Location.

M132071202LC_pyr

M132071202LC_pyr Enlargements

-H-187 (LO-III-012-H1) PLATE- 2, 1 L-BAR" is in square. Other "towers" in circles appear to be surrounding pyramidal structures to complement the "L". (Kopal, Pl.308, p.280)

-53 Tower sketches, perspective drawings with shadows.

LO-II-182-H1 (NASA 67-H-041) PLATE 27 Control Wheel in crater. (Drawing p. 179);

And 70-H -1630 Craters with wheels.

Leonard's sketches of Machine tools, T-bars and diamond shapes.

Rolling Stones: Moon As Viewed by Lunar Orbiter p. 97 and 119.

Crater Gruithuisen K. NASA 125-H1. Moo as viewed by LO p.69; and AS8-12-2052;

NASA 168-H3; and NASA 60-H2

LO-III-118-M, 118-M [2] and 118-H2.

67-H-935 (LO-IV-187-H2) (Leonard pl.29)

LO-IV.187-2 Medium shot of BLOB "dome" over crater rim; and close-up.

LO-IV-187-2 Enlargement; and LO-IV-186-H3 and 187-H2 "Black crosses

Extreme close-up of IV-186-H3 Pattern of "spheres"

PYRAMIDS of AS15-M-2087, 2088 and frame 2089

Leonard's drawing of the "domes" p.182

Domed habitat using crater for protection. ("Domed Cities." p. 4)

Regolith "lunar soil" covering dome for shielding purposes. ("Domed Cities..." p. 20)

ALPINE VALLEY construction site: 67-H-897 (LO-IV-115-H3) (See also LO-IV-H-116-H1)

LO-IV-115-H3 Dome, top left; 'sawhorse' dome top right.

Mare Crisium "bridge" on a clear night. (Compliments of Alex Norman, 'Anacortes Astronomy Club')

Mare Crisium Promontories at different times. (Compliments of http://www.the-moon.wikispaces.com/www.the-moon.wikispaces.com)

Mare Crisium "bridge." (Compliments of Alex E. Norman, 'Anacortes Astronomy Club')

Leonard's drawing, p.188

Sculptured "duck" and "skull" mounds.

AS16-121-19438 Apollo shot of northern rim of the Sea of Crisis - 74 SEA OF CRISIS: Pictures of the "bridge" LO-IV-1191-H3-81

Mare Crisium: "Bridge" as see by Astronomy club

AS15-94-12753 Close-up of Mare Crisium, Apollo 15 photo; and AS15-94-12752 (Another section of Mare Crisium)

IV-061-H2 and

IV-191-H3

NAC M113168034R "The Bridge" Another amazing bit of lunar geology revealed by LROC!

M130863593L/R

M113168034R

M113168034R

M123791947L. See also, (M103725084L, M103732241L, M106088433L, M113168034R, M123785162L, and M123791947L).

M126710873R [NASA/GSFC/Arizona State University] "Bullet Hole" and the "Pit" in Mare Tranquillitatis

"THE PIT" Pits in Mare Tranquillitatis; Mare Ingenii and Marius Hills (LROC)

AS16-M-2493_med Bullialdus-Lubinicky Area

Ibid, AS16-124-19901

AS16-M-2496_med

M119062083MC. King Crater on the far side of the Moon – LROC

-H-839 (AS16-120-19228). PLATE-12. Circles indicate several small craters in the process of being worked

-H-834 (AS16-120-19265) PLATE-10. X-drone w/no spraying

-H-836 (AS16-120-19226). PLATE-11. King Crater

Ibid, Gear teeth sticking out of rubble?

-94 Apollo-16 AS16-M-2496_med and Leonard's sketches of gears, teeth, 3-mile shaft and other unknowns.

Leonard's sketches of stitching, crater chains, LUBINICKY Area.

M119062083MC King Crater on the far side of the Moon – LROC

Apollo-16 photo 72-H-839 (AS16-120-19228) Leonard's PLATE-12.

72-H-834 (AS16-120-19265) PLATE-10. X-drone w/no spraying; and 72-H-836 (AS16-120-19226) PLATE-11. King Crater.

AS16-120-19228

72-H-834(a) (AS16-120-19226) PLATE-11. Top box shows the "sprayer" unit.

Leonard's attempt to draw an alien device in Crater King. Looks like a huge "cannon." (p.80)

LO-I-66-H-1293 "DINOSAUR CRATER" in Un-named crater on Far side of moon.

-106 Ibid. Images of mound icons "eagle", "dinosaur" and dot pattern; and sketch of X-drones.

CRATER KING AS16-M-0359_med Location of the X-Drones.

NASA As16-M-0359, 0890-M,

LO-I-136-H3. Unnamed Crater on far side of Moon. Nickname: "Icon or Dinosaur Crater"

NASA Photo 66-H-1293

Enlargement of LO-I-136-H3

AS16-M-1322 Leonard's 72-H-1113 photograph on page 141 is of Lobachevski Crater; a crater NW of King Crater on the far side of the Moon at 10 degrees North by 112 degrees West.

Lobachevski Crater photographed by Clementine

-113 Ibid. AS16-P-5029_FULL_MED; and frames 5022 and 5029

71-H-781 (AS14-70-9686 and 9688) PLATE 6 Leonard's "super-rig"

LROC frame M1100495581L. Enlargement of Pasteur D crater rim and "Big-Rig" area.

-H-8 (AS8-13-2244)

AS17-151-23115. Crater Doppler area.

Apollo-8 photo 69-H-8 (AS8-13-2244)

Drawing of crater "rope-ladder" p. 175

LO-I-036-H3

LO-II-037-H1 Letters, numbers, clamps and cables.

LO-II-108-H1

LO-II-162-H3 Just dirt and rocks! (LO-IRP enhancement: "MoonView.com")

-118 LO-II-162-H3 Just dirt and rocks! (LO-IRP enhancement: "MoonView.com"); and

AS10-33-4906; "Coffee Bean" Crater; and LO-III-213-H1-[2] Smoke Spiral Crater.

Snap_2011.07.13_11h25m39s_005. PYRAMID IN CRATER. (Supposedly shot by Hubble telescope, 2008)

AS15-M-1554_MED

AS16-121-19407 THE LOBACHOVSKY TOWER. Zoom-in on crater let impact point to what "looks" like a tower or standing object.

AS16-121-19407 and AS16-P-5024 Enlarged. Notice the rim is highlighted [1] and there is no mysterious object "tower" [2].

AS16-P-5024 Enlarged. Lobachevski impact craterlet enlarged. LROC Photo M176899195LC. Crater Lobachevski from a direct 'top view'.

AS16-121-19407

AS16-P-4559_FULL_MED. Crater Kant P. Impact crater landslide, rim collapse slide... La Pérouse A Crater IMPACT LANDSLIDE NAC mosaic (M152390311LR), image is ~4.5 km across [NASA/GSFC/Arizona State University].

Snap_2011.07.13_11h25m39s_005. PYRAMID IN CRATER. (Supposedly shot by Hubble telescope, 2008)

AS15-M-1554_MED

AS16-121-19407 THE LOBACHOVSKY TOWER. Zoom-in on crater let impact point to what "looks" like a tower or standing object.

AS16-121-19407 and AS16-P-5024 Enlarged. Notice the rim is highlighted [1] and there is no mysterious object "tower" [2].

AS16-P-5024 Enlarged. Lobachevski impact craterlet enlarged. LROC Photo M176899195LC. Crater Lobachevski from a direct 'top view'.

AS16-121-19407

AS16-P-4559_FULL_MED. Crater Kant P. Impact crater landslide, rim collapse slide... La Pérouse A Crater IMPACT LANDSLIDE NAC mosaic (M152390311LR), image is ~4.5 km across [NASA/GSFC/Arizona State University

Ibid. Craterlet LEFT: Enlargement as done by graphics expert. RIGHT: Enlargement of true angle – shows highlight at 45%

BIBLIOGRAPHY:

Bonnette, Dennis "Origin of the Human Species."

Burgess, F. and W. D. Whitley "Text Book of Hindu Astronomy."

Cortright, Edgar (1968) NASA SP-168 "Exploring Space with a Camera."
    NASA, Washington D.C.

DiPietro, Vincent and Gregory Molenaar (1982) "Unusual Martian Surface features."

El-Baz, Farouk and L.J. Kosofshy (1970) "The Moon as Viewed by Lunar Oribiter."
    NASA SP-200

Fielder, Gilbert (1965) "Lunar Geology." Lutterworth Press, London

Golden, Fred (April, 1985) "Discover"

Hoagland, Richard (1991 Fall Vol-1, No2) "Martian Horizons"

Hoagland, Richard (1991, Summer Vol-1) "Martian Horizons"

Hoagland, Richard (1992 Winter Vol-1, No.3) "Martian Horizons"

Hoagland, Richard (1994, Winter Vol-1, No.4) "Martian Horizons"

Hoagland, Richard (1995, Summer Vol-2, No.5) "Martian Horizons"

Kaysing, Bill "We Never Landed A Man on the Moon."
    and "We Never Went to the Moon."

Kopal, Zdenek (1960) "The Moon." Chapman and Hall, London

Kopal, Zdenek (1971) "A New Photographic Atlas of the Moon." Taplinger Pub. Co.

Leonard, George (1976) "Somebody Else is on the Moon" (David McKay, Pub.N.Y.)

Loomis, Alden (1965 Oct. Vol-76, No.10) "Some geological Problems of Mars."

Geo. Soc. Of America Bulletin

Midnight (Feb 8, 1977)

Moore, Patrick (1953) "Guide to the Moon."

More, Patrick (1950) "The Planet Mars."

Mutch, Thomas (1972) "Geology of the Moon A Stratigraphic View."

Princeton University Press, N.J.

NASA (1968) "The Mars Book." SP-179, 1968 Edition.

NASA (1974) "Mars as Viewed by Mariner 9." NASA SP-329

NASA Washington D.C.

NASA SP-206 "Lunar Orbiter Photographic Atlas of the Moon."

David Bowker and Kenrick Hughes (1971)

NASA SP-241 (1971) "Atlas and Gazetteer of the Near Side of the Moon."

Guischewski, Kinsler and Whitaker

National Enquirer (Oct 25, 1977)

Nelson (1955) "There Is Life On Mars."

Schultz, Peter (1976) "Moon Morpoholgy" Univ. Tex. Press

Shneour, Elie and Eric Ottensen (1966) "Extraterrestrial Life: An Anthology and

Bibliography" Nat. Acad. Of Sc.

Short, Nicholas (1975) "Planetary Geology." Prinston Hall, Inc.

Spurr, J. E. (1944) "Geology Applied Selenology." Science Press Printing Co.,

Lancaster, Penn.

Spurr, J. E. (1949) "Geology Applied to Selenology." Volume - IV,

Literary Licensing

Sreckling, Fred (1981) "We Descovered Alien Bases On The Moon."

(G. A. F. International)

Surya Siddhanta

Technology and Youth (May 1968)

Thomas, Andrew (1971) "We Are Not The First" London, Sphere.

Wallace, Alfred Russell (1907) Is Mars Inhabitable?"

SOURCES AND LINKS:

1.) FREE PDF of book: "Somebody Else is on the Moon." By George Leonard.
https://ia600404.us.archive.org/16/items/SomebodyElseIsOnTh eMoon/SomebodyElseIsOnTheMoon.pdf

2.) https://www.metabunk.org/threads/debunked-alien-base-on-the-moon-triangle-of-dots-photo-artifact.2965/

3.) http://www.themortonreport.com/discoveries/paranormal/aliens-on-the-moon/

4.) http://ufodigest.com/article/who-else-could-be-moon

5.) http://www.godlikeproductions.com/forum1/message218714/p g1

6.) http://www.paranormalnews.com/article.aspx?id=1185

7.) http://gizadeathstar.com/2011/05/the-idea-that-will-not-go-away-bases-on-the-moon/

8.) http://www.thescienceforum.com/pseudoscience/381-3-different-moon-images-same-location-different-objects.html#post527548

Literate world history took shape during the course of the third millennium BCE chiefly in the Mesopotamian land of Sumer.

There is a vast difference between the way secular scholars process this data and the way believers in the Bible can and should process it. By accepting at face value both the chronological perspective of the Bible and the high longevities of the Noahic patriarchs, biblicists can make sense of Sumerian data and revolutionize the image of world history at its source. To make good on this premise, it is essential to compare and match names from king lists and mythological pantheons. What emerges from these comparisons is a set of fifty-four feudal and imperial aristocrats who created world civilization in their own image.

Once these persons are known, world history loses its aura of randomness and anonymity and takes shape as a single, variously detailed story. This book details the genealogical comparisons of all the nations mythological pantheons with the Genesis Chapter 10 list of post flood patriarchs and establishes a foundation for building a true history of mankind. CreateSpace eStore:

https://www.createspace.com/5579211

7.) THE SALVATION OF ALL: Fulfilling the Restoration of All (Acts 3:21). by Ross S Marshall. 94 pages. Every human being ever born shall see the salvation of God. For, all flesh was made by God and for God, and this was very good - Gen.1:27. https://www.createspace.com/5543783

8.) GOD'S HELL NOT FOREVER: All Hell Breaks Loose! (Rev. 20:13-14). by Ross S Marshall 102 pages. https://www.createspace.com/5543933

9.) GOD'S LOVE WINS ALL: God's Love does not loose any! by Ross S Marshall. 124 pages. Jesus declared, "he [his Father] is kind unto the unthankful and to the evil" - Lk. 6:35.
https://www.createspace.com/5543834

10.) UNIVERSAL SALVATION: 126 pages. By R. Marshall. All verses show God saves all humans in the end.
https://www.createspace.com/5534312

1.) "STELLAR STERILITY": The Search for Alien Presence in the Universe. by Ross S Marshall 590 pages.
https://www.createspace.com/4958613

2.) "IS ANYONE ELSE ON THE MOON?" By R.S. Marshall. Vol-1

Continuation of Search for Alien Artifacts on the moon. 340 pages.
https://www.createspace.com/4676724

3.) "MOON MARS MONUMENTS MADNESS." By R.S. Marshall. Vol-2. Study in Richard Hoagland's Moon Mars Alien Artifacts & Fred Steckling's Alien Bases. 232 pages.
https://www.createspace.com/4724107

5.) "NASA TECHNICAL PUBLICATIONS" Vol-1: THE MARTIAN METEORITES: The Search for Extraterrestrial Life.
  by Ross S Marshall 486 pages.
   https://www.createspace.com/5281579

6.) "THE ALL MANKIND BIBLE COMMENTARY": A Study of Universal Reconciliation of All Mankind. by R.S. Marshall
This book brings to light the positive "good news" of the message as trumpeted by the angel of the Lord, of the universal reconciliation of all mankind and the whole of creation.

Thomas Talbott said,...
"If I am ignorant of, or deceived about the true consequences of my choices, then I am in no position to embrace those consequences freely; and similarly, if I suffer from an illusion that conceals from me the true nature of God, or the true import of union with God, then I am again
in no position to reject God freely" (The Inescapable Love of God, p. 187).
Similarly, if I am enslaved to my destructive desires and passions,
then I am not in a position to make a free decision. Just as addicts are incapable of making free and responsible decisions until they have secured liberation from the drugs that enslave them, so those who are in bondage to their passions are incapable, to the degree they are so bound, of free decisions and actions—they could not have done otherwise."
We must rethink the true definition of human free will
and understand that it is weak compared to God Omnipotent power
and will. Human's break down, rethink, and do change under pressure
  - God does not. Man's "will" cannot in any way pressure God's will to fall short in "willing all men to be saved...".
https://www.createspace.com/4949875

MY YOUTUBE:
http://www.youtube.com/watch?v=1RKfyTVMgvE&list=HL1385879974&feature=mh_lolz